Soil Cover of the Dried Aral Seabed in Kazakhstan

Gulnura Issanova • Jilili Abuduwaili
Kuralay Tynybayeva

Soil Cover of the Dried Aral Seabed in Kazakhstan

Springer

Gulnura Issanova
Research Center for Ecology and
Environment of Central Asia
Almaty, Kazakhstan

Jilili Abuduwaili
Xinjiang Institute of Ecology and
Geography, Chinese Academy of Sciences
Urumqi, Xinjiang, China

Kuralay Tynybayeva
U.U. Uspanov Kazakh Research Institute of
Soil Science and Agrochemistry
Almaty, Kazakhstan

ISBN 978-3-031-29866-0 ISBN 978-3-031-29867-7 (eBook)
https://doi.org/10.1007/978-3-031-29867-7

This Springer imprint is published by the registered company Springer Nature Switzerland AG
The registered company address is: Gewerbestrasse 11, 6330 Cham, Switzerland

Preface

This book considers and contains research and fieldwork results on soil cover in the dried bed of the Aral Sea, Kazakhstan. The dried seabed of the Aral Sea is a newly formed surface and is an anthropogenic salty desert (Aralkum) now. It is an ecologically catastrophic region in Central Asia particularly in Kazakhstan.

The shrinking of the Aral Sea awakened a huge chain of ecological processes that led to the emergence of the Aralkum desert as a result of soil and vegetation degradation. Soil salinization is among the most well-known land degradation processes determining the ecological state of the land surface.

The aim of the study was to conduct a comprehensive analyses and assessment of the current state of the soil cover in the dried Aral Seabed in Kazakhstan. Namely, morphological characteristics of the soils and their chemical-physical properties were studied in order to know the material properties can be transported during soil deflation as dust/sand/salt storms in the region. Soil cover in the dried Aral Seabed was mapped and the map were updated and modified with the result of fieldwork data and satellite images with remote sensing indexes. As well as the changes in long-term dynamics of climatic parameters (temperature, precipitation, and evapotranspiration) in the Aral Sea region were revealed and aridization process in general.

In addition, processes that are occurring in the dried Seabed soils as a result of anthropogenic impacts inducing soil degradation and desertification (soil salinization and deflation) were studied.

This study has a great importance in helping to predict and monitor the process of soil salinization, issues of halogenesis and migration of salts in the pedosphere and atmosphere during storms. The study results can be used in development measures to combat saline desertification and combating sand moving processes in the dried Aral Seabed.

The book is mainly addressed to scientists and researchers whose research has been focused on land degradation and desertification studies, as well as students and planners. It is intended to be a source of information and inspiration for all readers who feel responsible for initiating the sustainable development and sustainable use of natural resources in Central Asian countries.

We believe that this publication provides a contribution to our knowledge about the soil properties in the dried Aral Seabed, their spatial distribution, and the environmental issues they create.

Almaty, Kazakhstan
Urumqi, China

Gulnura Issanova
Jilili Abuduwaili
Kuralay Tynybayeva

Content and Structure of the Book

This book summarizes the outcomes of research results based on field works conducted in the dried Aral Seabed (Aralkum desert) and recent studies related to soil cover of the dried Aral Seabed and its region (Kazakhstan). The book has six individual chapters as follows:

Chapter 1, "Introduction and Background on Environmental Changes in the Dried Aral Sea Region," provides an overview on studies of environmental issues (soil salinization, deflation, desertification) brought by desiccation of the Aral Sea. As well as data sources and research methods used in the study of the soil cover in Aralkum desert were generalized. Chapter 2, "Environmental Conditions of the Aral Sea Region," considers information about geographical location of the sea, geological structure, climatic conditions and climate change, vegetation cover and its degradation.

Chapter 3, "Origin of the Aralkum Desert (Dried Aral Seabed)," considers research results on sea water surface area shrinkage and dust and sand particle transportation.

Chapter 4, "Soil Cover of the Dried Aral Seabed (Aralkum)," provides results about morphological properties of the soils, their chemical compositions, salts in the soils and salt profiles, heavy metals in the soils, and soil texture.

Chapter 5, "Formation, Degradation and Mapping," considers information on soil forming processes, soil types and soil transformation in soil properties and types, soil degradation processes, and soil maps of the Seabed (1980s, 2020s), and Chap. 6, "Conclusions" and Chap. 7, "Recommendations for the Rational Use of the Land and Water Resources in the Dried Aral Seabed Region".

Acknowledgments

Certain scientists and researchers have supported and provided the basis for this book.

We are thankful to doctor in Biological Sciences, Konstantin Pachikin for his useful ideas and experiences.

Special thanks go to Azamat Kaldybayev (PhD) and Aigerim Kalybayeva (MSc), whose help made this work possible for us by mapping the soil cover for the analysis part in soil type's distribution.

We thank Amandyk Saduakhas for his assistance with soil sampling during fieldwork and providing some basic and essential materials as illustrative materials to accomplish this uphill task. As well as we thank Kanat Kulymbet (MSc) for making analysis in vegetation cover of the dried Aral Seabed in Kazakhstan part.

We sincerely thank all of the scientists and researchers, in particular Dr. Victor Dukhovnyi and Dr. Philip Micklin, whose contribution in study and research of the soil cover in the dried bed of Aral Sea and water resources of the Aral Sea basin became part of this book.

We also thank our friends, colleagues and well-wishers for their precious and continued support, love and facilitation on many different occasions during the writing of this book.

This research was funded and supported by the Strategic Priority Research Program of Chinese Academy of Sciences (Grant No. XDA2006030102), the National Natural Science Foundation of China (Grant No. 42171014), the Chinese Academy of Sciences President's International Fellowship Initiative (2020PC0090), the Foundation of State Key Laboratory of Desert and Oasis Ecology, Xinjiang Institute of Ecology and Geography, Chinese Academy of Sciences, and Science Committee of the Ministry of Education and Science of Kazakhstan (Grant No. AP08053270: "Assessment of the current state of the soil cover of the Eastern Aral Sea region and its transformation as a result of the drying of the Aral Sea").

Contents

About the Authors

Gulnura Issanova holds a doctorate degree in Natural Sciences from Xinjiang Institute of Ecology and Geography, Chinese Academy of Sciences (China), is an associate professor in the Al-Farabi Kazakh National University, scientist and researcher at U.U. Uspanov Kazakh Research Institute of Soil Science and Agrochemistry, and a scientific secretary at the Research Centre for Ecology and Environment of Central Asia (Almaty), Kazakhstan. Her research interest is focused on problems of soil degradation and desertification. She participates regularly in the International Scientific Activities on Environmental Problems as well as writes articles on the subject and takes part in local and international projects. She has published more than 70 articles in international and domestic journals and 6 books: *Aeolian Processes as Dust Storms in the Deserts of Central Asia and Kazakhstan* (2017), *Man-Made Ecology of East Kazakhstan* (2018), *Hydrology and Limnology of Central Asia* (2019), *Geomechanical Processes and Their Assessment in the Rock Massifs in Central Kazakhstan* (2019), *Current and Future Trends of Rainfall Erosivity and Soil Erosion in Central Asia* (2021), and *Tourism in the Kyrgyz Republic* (2021) by Springer Nature. Gulnura Issanova became a Laureate of the International Award "*Springer Top Author*" and awarded in the Nomination "Springer Young Scientist Awards" for high publication activity in scientific journals published by Springer Nature, 2016; the *best young scientist* of the Al-Farabi Kazakh National University, 2019; *Owner* of the State scientific scholarship for *talented young scientists* by the Ministry of Education and Science of Kazakhstan (2019–2020); and *Winner* of the personal scholarship of the President of the Chinese Academy of Sciences (PIFI) for Postdoctoral Researcher (2020–2022).

Jilili Abuduwaili is a professor and deputy director at the Xinjiang Institute of Ecology and Geography, Chinese Academy of Sciences (China) and an Executive Director of the Research Center of Ecology and Environment of Central Asia (Almaty). He holds a doctoral degree from the Lomonosov Moscow State University, Russia. He is a member of a council of the China Natural Resources Society and the Xinjiang Geographical Society as well as a deputy chairman of the Xinjiang Natural Resources Society. Jilili Abuduwaili is an associate editor of the *Journal of Arid Land* and editor-in-chief of the journals *Arid Land Geography* and *Arid Land Research* (both in Uyghur language). He is also an editorial board member of the journals *Arid Ecosystem* in Russia and *Soil Science and Agrochemistry* in Kazakhstan. He is a Supervisor for postgraduate students. Since 2015, he has been a foreign academician at the Academy of Agriculture of Kazakhstan. His main fields of work and expertise are in the areas of natural resources and environments, ecological safety, lake environments, disaster, and landscape changes in arid lands. Jilili Abuduwaili has built international cooperation in research between Russia and Central Asian countries. He has published more than 200 papers in international and domestic journals and 7 books: *Lakes and Salt Dust Storms in Arid Lands* (in Chinese), *Overview of Central Asian Environments* (in Chinese), *Climate Change Effects to the Ecological System in TianShan Mountains* (in Russian), *Natural Geography of Kyrgyzstan* (in Russian), *Water Resources and Utilization in Uzbekistan* (in Russian), *Arid Soil in Kazakhstan: Current States and Utilization* (in Russian), and *Salt Dust Storms: Special Disaster in Arid Lands* (in Uyghur).

Kuralay Tynybayeva holds a PhD in Soil Science and Agrochemistry. She is a researcher in Soil Ecology department at the U.U. Uspanov Kazakh Research Institute of Soil Science and Agrochemistry. She regularly takes part in scientific internships (University of Putra, Malaysia, Lomonosov Moscow State University, Russia) and in research projects dealing with soil salinization in the desert zone of Kazakhstan. Her scientific interests cover a wide spectrum, as she participated in various projects that investigated the soils of different regions in Kazakhstan. She has published over 20 scientific articles related to her studies.

Abbreviations

CA	Central Asia
HM	Heavy metals
TLV	Threshold limit value
WS	Weather station

Chapter 1
Introduction and Background on Environmental Changes in the Dried Aral Sea Region

Abstract This chapter includes an introduction and a literature review on researches of soil cover and environmental changes in the dried Aral Sea region. Since the Aral Sea region is one of the ecological regions in Central Asian countries and it brings environmental problems such as soil deflation (dust/sand/salt storms), soil salinization and intensive desertification, and social-economic issues, it became one of the important and relevant topics of research community not only in Central Asia but also of the world. Furthermore, data sources on soil cover and research methods for the study were provided. Field soil sampling process and lab chemical-physical analysis of soils, cartographic materials, and satellite images used in the research were proposed and discussed.

Keywords Aral Sea · Hydrological regime · Desiccation · Environmental problems · Central Asia

1.1 Background of Research on Soil Cover and Environmental Changes in the Aral Sea and Its Region

The Aral Sea was located within Uzbekistan and Kazakhstan in the northern desert part of Central Asia. The Aral Sea basin has a complex structure. During the period of sustainable existence of the Aral Sea ecosystem, the sea level was maintained at 53 m, which is almost 80 m higher than the level of the Caspian Sea. The width of the sea along the 45° parallel was 265 km, and the length of the coastline exceeded 4430 km. The area of the water surface of the Aral Sea before the fall of its level, in the 1960s, was 69.79 thousand km^2. The maximum depth of the sea was 69 m, and the volume of the water mass was about 1056 km^3 (Dukhovnyi et al. 2008).

The hydrological regime of the Aral Sea, as well as most drainless water bodies of the arid zone, is subject to significant fluctuations under the influence of natural and anthropogenic factors. Conducted in the area of the Aral Sea, geological,

G. Issanova et al., *Soil Cover of the Dried Aral Seabed in Kazakhstan*,
https://doi.org/10.1007/978-3-031-29867-7_1

geomorphological, and archaeological studies have shown that in the previous 4000–6000 years, the amplitude of fluctuations in its level was more than 20 m (Bogdanov and Pinkhasov 1995; Kes 1969; Lymarev 1967; Pinkhasov 1998; Rubanov 1987). Fluctuations in the level of the Aral occurred before 1960; however, their amplitude over the past 200 years did not exceed 4 m, and it was 1 m in the first half of the last century. Until the 1950s, the ecological situation in the region was quite stable (Dukhovnyi et al. 2008; Micklin 2016).

The Aral Sea is an ecological catastrophic crisis, which occurred since 1960s, and is expressed in the most acute social and environmental problems, including soil salinization and desertification, degradation of vegetation (Issanova et al. 2022), climate change (Lioubimtseva 2014), progression of various diseases among population of the region (Anchita et al. 2021), and other negative processes.

Since 1960s, the process of development of new irrigated lands began in the Central Asian region, which led to the intensive use of the river flow of two Syrdarya and Amudarya rivers. To make matters worse, agricultural production was concentrated on growing very moisture-loving crops, mostly monocultures of cotton and wheat. This was the main reason for the drying up of the Aral Sea. Since this period, there has been a steady decrease in the sea level. Due to the rather shallow nature of the sea, a decrease in the volume and level of water has led to a significant change in the configuration of the coastline and a reduction in the area of the water surface and an increase in the desert adjacent to the Aral Sea.

The fall in the sea level has accelerated markedly since the middle of 1970s, due to the removal of natural river flow for irrigation. From 1975 to 1980, the sea level fell already by 3.26 m, and the sea level decreased by 0.65 m per year. The most intense level drop of the sea was observed when the Amudarya river did not reach the Aral Sea during 1973–1990 (Fig. 1.1).

The main consequences of the drying up of the Aral Sea, in addition to a decrease in the volume and area of the water surface and growth and changes in the nature of its mineralization, can be confidently attributed to the formation of a dried seabed in place of a huge salt desert with an area of almost five million hectares by now. As a result, a unique freshwater reservoir gave way to a huge bitter-salty lake in combination with a colossal salty desert at the junction of three sandy deserts (Dukhovnyi et al. 2008; Micklin 2010).

In the result of sea drying, the Aralkum sandy-salt desert formed with almost four million ha territory in the former seabed and only halophytes can be grown in this area (Matzhanova 2007; Breckle 2013). The resulting desert intensifies the processes of desertification of landscapes around the former sea because the rapid drying of the Aral Sea has led to global changes in all natural conditions, including changes in soil cover (Stulina and Sektimenko 2004; Issanova et al. 2022) and climate (Lioubimtseva 2014).

The desiccation of the sea has become an ecological disaster in whole Central Asia (Breckle 2013). Drying and aridization of the vast territory of the Aral Sea region is accompanied by the cessation of floods and water channels, the widespread occurrence of solonchaks, a decrease in land productivity and soil fertility, and their sanding. Almost 90% of the dried Aral seabed has undergone

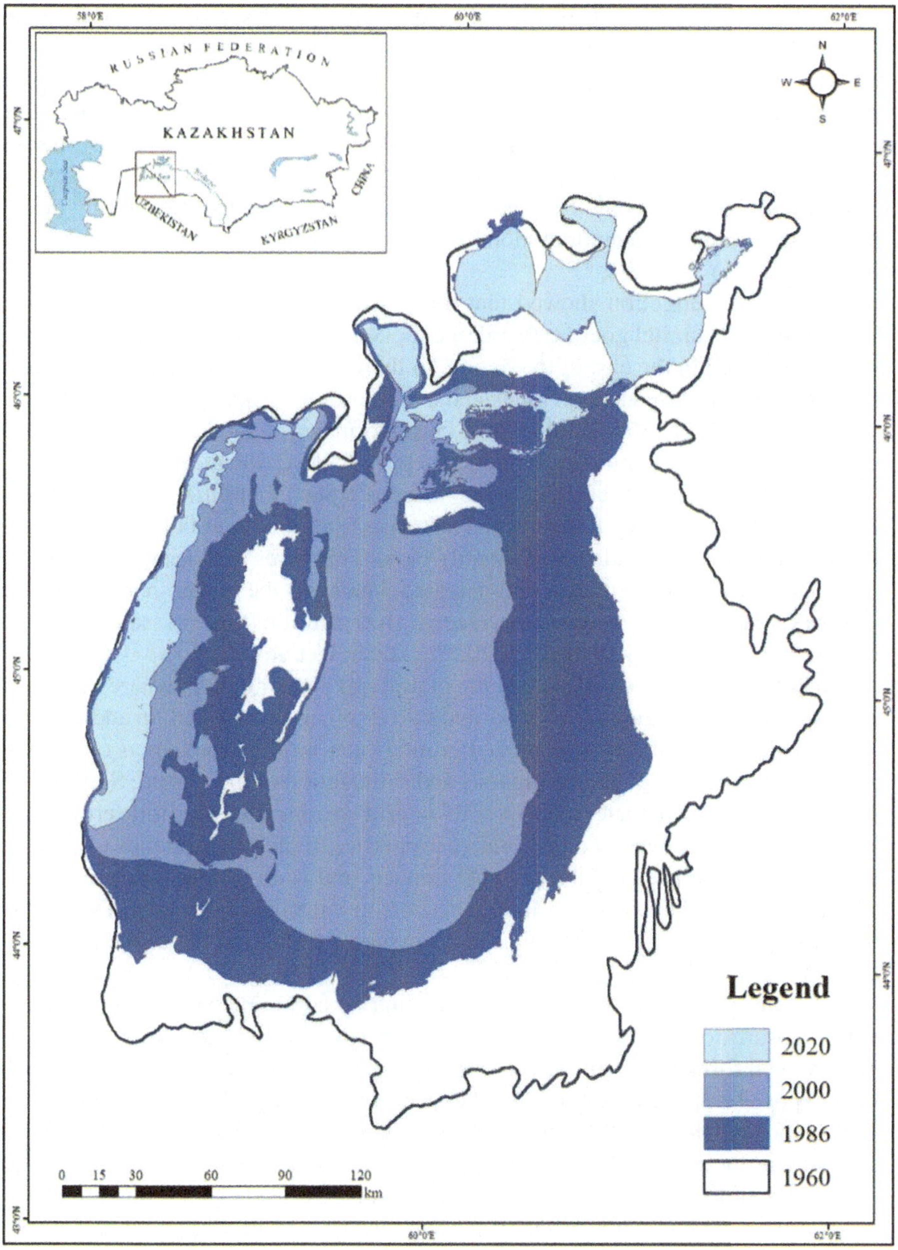

Fig. 1.1 Changes in the level of the Aral Sea from 1960 to 2020

desertification process. According to researches conducted in 2007–2008 on the territory of the eastern and the southeastern part of the dried Aral seabed, in conditions of aridization, territories of transformed soils were exposed to desertification and wind erosion and, in areas with soils of light mechanical composition, the development of sanding of soils, dust transfer, and salt accumulation (Tomina 2009a, b). The dried part of the seabed, especially eastern and northern, have become the main sources of large dust storms and centers of dust and salt removal to the surrounding area (Dukhovnyi et al. 2008; Semenov 1995, 2009). They are located mainly in soils with a poorly developed profile on loose solonchaks and sandy massifs. The activity of deflation processes is almost year-round; however, it is most intensively developed in the summer and autumn months and is associated with a period of drought (Semenov 2011).

Sand transfer direction showed that the movement of massifs of mobile sands north of the 46° parallel goes eastward; the movement of sands located south of this parallel observed in the east of the region, in the western direction; and in the areas adjacent to the Aral Sea and on the islands, in the southwestern and southern direction (Semenov 2011; Shen et al. 2016). Dust storms, droughts, and dry and strong winds are the most significant climatic factors of desertification (Issanova et al. 2014; Semenov 2011). According to the national report, 75 million tons of sand, as well as 65 million tons of fine dust and salt, rise from the dried bottom of the Aral Sea, where winds and cyclones constantly blow. This leads to a decrease in pasture productivity and crop yields. The irrigated lands of the lower reaches of the Amudarya and Syrdarya annually receive 100 million tons of salt and sand (Nauryzbayeva 2019; Aiimbetova 2020; Ramazonov et al. 2020).

The drying Aral seabed is an example of arid salt accumulation, where the different nature of the emerging solonchaks and saline soils is manifested. In addition, the close occurrence of highly mineralized groundwater on the beach areas of the drying seabed contributes to the continuous and salinization of the seabed. Soil salinity is one of the main characteristics of soils in arid regions, limiting soil fertility and determining their genetic and meliorative properties. It quickly responds to changes in natural conditions and anthropogenic impacts and is an informative property of the state of soils. Therefore, in arid regions, it is monitoring of changes in soil salinity that makes it possible to obtain the necessary information about the processes taking place in soils.

Soil is an indicator of natural processes, and its condition is the result of long-term exposure to various sources of pollution and deterioration. The process of primitive soil formation in the seabed began on various sandy deposits, sandy loamy rocks exposed on a dried bottom (Orazbekova and Budnikova 2014; Wu et al. 2021). These deposits with a light-textured composition led to an increase in deflationary processes in the region. As a result, saline dust was introduced into the Aral Sea region, including irrigated lands. Soil deflation process as dust and salt storms was noted with periodic removal of sandy and salt materials from all types of solonchaks in the dried seabed (Issanova et al. 2015; Zucca et al. 2021).

The degradation of the resource potential of the Aral Sea has taken an irreversible direction, and the preservation and restoration of which in the foreseeable future is practically impossible. A catastrophic drop in sea level, a decrease in the volume of water mass, an increase in salinity led to the loss of its biological diversity. Changes in the basic chemical and physical characteristics of soils and habitats of aquatic organisms led to a reduction in the number of microorganisms in the small and big Aral Seas.

1.2 Data Sources on Soil Cover and Research Methods

Field Soil Sampling

Fieldwork was carried out in June–July 2020 and 2021 (Fig. 1.2a–c) on the soils of the dried bed in the northeastern and eastern part of the Aral Sea in Kazakhstan (Fig. 1.3). During the fieldwork, 41 soil profiles were made and 0–160 cm in depth were taken, and 204 soil samples were taken from all those soil profiles for analysis of the physicochemical, chemical, and morphological properties of the soil.

All these soil samples were grouped into 10 salt profiles in 2020 and 10 soil types in 2021, according to the morphological characteristics and chemical properties of the soils. The coordinates of the soil profiles were also obtained and added to the compiled map (Fig. 1.3).

A morphological description of the soil profiles was carried out by genetic soil horizons visually. As well as the vegetation cover of the area, genera, and species of plants (visual projective vegetation cover and condition) were determined.

Chemical Lab Analysis of the Soils

The chemical (salts, CO_2, pH, heavy metals) and physical (soil texture, organic matter) properties of soils were determined: The content of salts according to the full composition of the water extract, i.e., its anions (CO_3^{2-}, HCO_3, Cl^-, SO_4^-) and cations (Ca^{2+}, Mg^{2+}, Na^+, K^+), and the content of carbonates (CO_2) were identified according to GOST 26425-85. The humus (organic matter) content (GOST 23740-79) as well as the content of heavy metals (Cu, Zn, Pb, Cd, Ni, Co, Mn, etc.) were identified according to GOST 27262-87, on atomic absorption spectrometer (Contr Aa300). The pH of the aqueous extract was measured by automatic potentiometric titrator "Compact model G20."

The electrical conductivity (EC) and Total Dissolved Solids (TDS) of soils were measured by ADDSJ-308A conductometer. The argentometric method (GOST 26425-85) was used to determine the cation-anion composition of (Cl^-), HCO^{3-}, and CO_3^{2-} with titration of sulfuric acid (GOST 26424-85), Ca^{2+} and Mg^{2+} based on complexometric method (GOST 26428-85), Na^+ and K^+ with flame photometry Contr Aa300 (GOST 26427-85), and SO_4^{2-} with sedimentation from a $BaCl_2$ solution (GOST 26426-112).

Fig. 1.2 Taking soil samples from the dried Aral seabed in 2020–2021

Climate Data Mapping

From the CHELSA climate database (https://chelsaclimate.org/), the average monthly temperature (T_{mean}), average monthly precipitation (Pr), and average monthly potential evaporation (PET) for 1986, 2000, and 2020 were used to

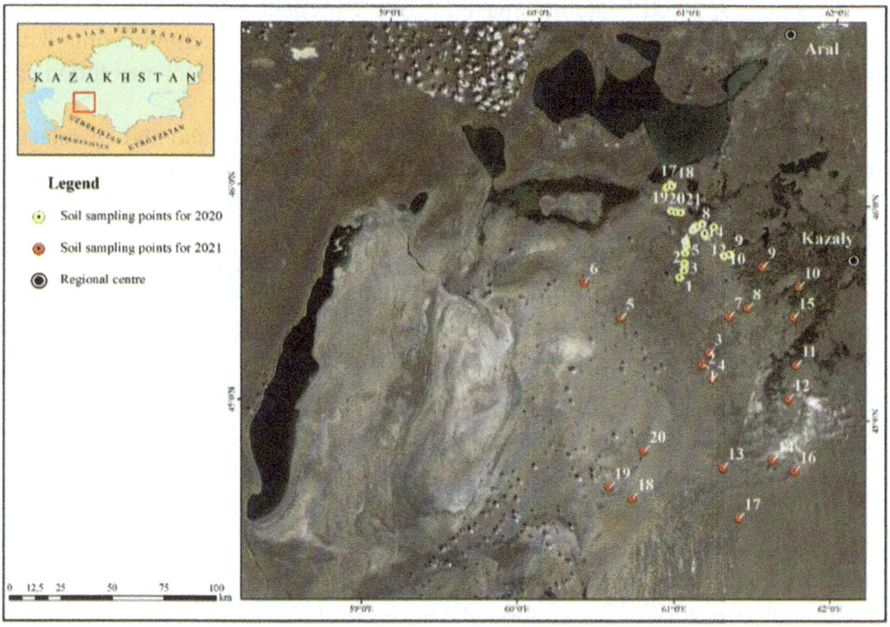

Fig. 1.3 Study area – dried Aral seabed

monitor the temporal and spatial changes in the climatic conditions of the northeastern Aral Sea region. As a result, climate maps were compiled for 1986, 2000, and 2020. Maps included climatic layers for various time periods and variables. Climate maps were compiled in ArcGIS 10.8 software.

The downloaded climate raster data were cropped and merged to obtain averages using the spatial analyst tools. The CHELSA database is based on mechanical statistical downscaling of global reanalysis data or global circulation model output.

Cartographic Materials and Satellite Data

All cartographic materials, including a soil map of Kazakhstan (2010; 1: 2500000), a map of soil salinity chemistry types in the USSR (1975; 1: 2500000), and a soil and vegetation degradation maps (2010; 1:3000000) were taken from the National Atlas of Kazakhstan (NAKZ 2010) as initial input data to generate maps using the ArcGIS application as the main tool. Furthermore, satellite images (Landsat 8, 10), the results of fieldwork data, and archival literary materials on the general properties of soils and their spatial distribution in the Kyzylorda region were collected/systematized and generalized in order to compile an updated and modified maps on soil cover of the dried Aral seabed in Kazakhstan part.

A Landsat 8, 10 OLI satellite image Level 1 T product with radiometric correction and co-registration with map projection and elevation corrections acquired during the growing season for June–July 2020 and 2021 at 30 m resolution was downloaded and applied.

In addition, according to the analyzed data on soil profiles and soil characteristics of the Modified Soil-Adjusted Vegetation Index (MSAVI), adjustments were made to the soil map. The updated soil map represents the spatial distribution of soil types and its degradation in the dried seabed.

MSAVI. The extent of land cover change can be estimated using the MSAVI and proposed a Modified Soil-Adjusted Vegetation Index (MSAVI) to lift limits on applying normalized difference vegetation index (NDVI) to the areas with a high composition of bare soil (Richardson and Wiegand 1977). The value of MSAVI ranges from −1 to +0.65. It can be expressed as the following:

$$\text{MSAVI2} = \frac{\left(2*\text{NIR}+1-\sqrt{\left(2*\text{NIR}+1\right)^{2}-8*\left(\text{NIR}-\text{RED}\right)}\right)}{2} \tag{1.1}$$

MSAVI was calculated as the ratio between R and NIR values.

In Landsat 4–7 MSAVI = (2 * Band 4 + 1 – sqrt ((2 * Band 4 + 1)2 – 8 * (Bans 4 – Band 3)))/2.

In Landsat 8 MSAVI = (2 * Band 5 + 1 – sqrt ((2 * Band 5 + 1)2 – 8 * (Band 5 – Band 4)))/2.

Distinguishing the output classes of vegetation cover was performed by subtracting pixels from the results of the strip algorithm with certain spectral characteristics of the same pixels in satellite images. After cross-checking the classification, a classification map was generated using valid data collected in the field. Cartographic materials were processed using ArcGIS® 10.8 software.

The MSAVI is mainly used for drought monitoring and soil erosion analyses. In addition, it is useful when analyzing plant growth, desertification, and pasture yields (Binte Mostafz et al. 2021).

The Normalized Difference Salinity Index (NDSI) was calculated using the spatial analysis tools in ArcGIS 10.8 software. The NDSI value fluctuates between −0.64 and 0.48.

As the salt content in the soil surface increases, the reflectivity at different wavelengths also increases, which is the basis for soil salinity monitoring using remote sensing method. The NDSI was calculated from the ratio of the difference between the red and NIR bands divided by the sum of two (Khan et al. 2005).

$$\text{NDSI} = \left(\text{RED}-\text{NIR}\right)/\left(\text{RED}+\text{NIR}\right)$$

RED – red range.
NIR – near infrared.

References

Aiimbetova S (2020) Environmental monitoring of soil salinization in the southern Aral Sea region. Eurasian Union Sci 9-78:10–12. (in Russian)

Anchita ZA, Khaibullina Z, Kabiyev Y, Persson K, Tussupova K (2021) Health impact of drying Aral Sea: one health and socio-economical approach. Water 13:3196. https://doi.org/10.3390/w13223196

Binte Mostafz R, Noguchi R, Ahamed T (2021) Agricultural land suitability assessment using satellite remote sensing-derived soil vegetation indices. Land 10:223

Bogdanov LA, Pinkhasov BI (1995) Report on ecological-hydrogeological and engineering-geological survey of the dried part of the Aral seabed (scale 1:200 000). Tashkent, p 253. (in Russian)

Breckle S-W (2013) From Aral Sea to Aralkum: an ecological disaster or halophytes' paradise. In: Progress in botany. Springer, Berlin, pp 351–398

Dukhovnyi VA (eds) (2008) Integrated remote sensing and ground studies of the dried bed of the Aral Sea. Tashkent, p 190 (in Russian)

Issanova G, Abuduwaili J, Semenov O (2014) Deflation processes and their role in desertification of the southern pre-Balkhash deserts. Arab J Geosci 7:4513–4521

Issanova G, Abuduwaili J, Galayeva O, Semenov O, Bazarbayeva T (2015) Aeolian transportation of sand and dust in the Aral Sea region. Int J Sci Technol 12:3213–3224

Issanova G, Abuduwaili J, Tynybayeva K, Kalybayeva A, Saduakhas A, Kulymbet K, Kaldybayev A, Erlan G, Tanirbergenov S (2022) Soil salinisation as a land degradation process in the dried bed of the North-eastern Aral Sea, Kazakhstan. Arab J of Geosci 15:1055. https://doi.org/10.1007/s12517-022-09627-w

Kes AS (1969) The main stages of the development of the Aral Sea. In: The Aral Sea Problem. Nauka, Moscow, pp 41–47. (in Russian)

Khan NM, Rastoskuev VV, Sato Y, Shiozawa S (2005) Assessment of hydrosaline land degradation by using a simple approach of remote sensing indicators. Agric Water Manag 77:96–109

Lioubimtseva E (2014) Impact of climate change on the Aral Sea and its basin. In: Micklin P et al (eds) The Aral Sea, Springer earth system sciences. Springer, Berlin/Heidelberg. https://doi.org/10.1007/978-3-642-02356-9_17

Lymarev VI (1967) The shores of the Aral Sea – an inland water body of the arid zone. Nauka, Leningrad, p 252. (in Russian)

Matzhanova (2007) Haloadaptation and ecological characteristics of plants in the southern dried bottom of the Aral Sea part due to desertification and soil salinization. Arid Ecosyst 13(33–34):113–117. (in Russian)

Micklin PP (2010) The past, present, and future Aral Sea. Lakes Reserv Res Manag 15:193–213

Micklin P (2016) The future Aral Sea: hope and despai. Environ Earth Sci 75:844

National Atlas of Kazakhstan (NAKZ) (2010) Environment and Ecology, vol 3. NAKZ, Almaty. (in Russian)

Nauryzbayeva Z (2019) Problems of soil formation and salinization in the southern Aral Sea region. Eurasian Union Scientists 9–66:4–5. (in Russian)

Orazbekova K, Budnikova T (2014) Environmental assessment of natural complexes of the dried bottom of the Aral Sea in the territory of Kazakhstan for the purpose of prospective development. Eurasian Union Sci 5:48–52. (in Russian)

Pinkhasov BI (1998) Forecast of the geoecological development of the Aral Sea. J Problems Desert Dev 3-4:179. (in Russian)

Ramazonov B, Faiziev V, Mutalov K, Zakirov D (2020) Desertification and transformation soil cover in the Aral Sea region. Acad Res Educ Sci 1(2):213–220. (in Russian)

Richardson AJ, Wiegand CL (1977) Distinguishing vegetation from soil background information. Photogramm Eng Remote Sens 43(12):1541–1552

Rubanov IV (1987) Geology of the Aral Sea. Fan, Tashkent, p 180. (in Russian)

Semenov OE (1995) Assessment of the scale of the removal mass of the Aral Sea aerosol. J Hydrometeorol Ecol 1:117–129. (in Russian)

Semenov OE (2009) Physico-statistical modeling of carrying sandy salt aerosol during storms in the Aral Sea region. In: Marburg international dust & sand storm (DSS) symposium "DSS and desertification", Marburg, pp 30–32

Semenov OE (2011) Introduction to experimental meteorology and climatology of the sand storms, Volkova, Almaty, p 580. (in Russian)

Shen H, Abuduwaili J, Samat A, Ma L (2016) A review on the research of modern aeolian dust in Central Asia. Arab J Geosci 9:625

Stulina G, Sektimenko V (2004) The change in soil cover on the exposed bed of the Aral Sea. J Marine Syst 47:121–125

Tomina TK (2009a) Soils of the dried Aral seabed. J Hydrometeorol Ecol 1:60–75. (in Russian)

Tomina TK (2009b) Erosion-deflation processes on the dried Aral seabed. J Soil Sci 2:1–6. (in Russian)

Wu Q, Yue H, Liu Y, Hou E (2021) Geospatial quantitative analysis of the Aral Sea shoreline changes using RS and GIS techniques. Earth Sci Inf. https://doi.org/10.1007/s12145-021-00714-2

Zucca C, Middleton N, Kang U, Liniger H (2021) Shrinking water bodies as hotspots of sand and dust storms: the role of land degradation and sustainable soil and water management. Catena 207:105669. https://doi.org/10.1016/j.catena.2021.105669

Chapter 2
Environmental Conditions of the Aral Sea Region

Abstract This chapter describes and analyzes environmental conditions (geographical location, geological structure, surface and groundwaters, climate and its change, vegetation cover and its degradation) of the dried Aral Sea region. The former fourth largest lake in the world is now a man-made and salty desert – Aralkum – with approximate area of 2.6 million hectares. The Aral Sea region is located in the Syrdarya depression and formed in the Upper Paleozoic with sediments of the Cretaceous, Paleogene, Neogene, and quaternary continental deposits. Syrdarya and Amudarya rivers are determined of the water balance of the sea. Changes in weather conditions were observed: the mean monthly temperature and potential evaporation in the region have increased by 2 °C (23.81%) and 76 kg m^{-2} (7.81%), respectively, from 1986 to 2020. The vegetation cover of the dried Aral seabed is covered mainly with shrubs, halophytes, and psammophytes (perennial saltwort plants, black and mixed saxauls). It has been degraded to strong (7%) and significant (21.2%) degrees.

Keywords Geographical location · Climate change · Vegetation degradation · Man-made desert

2.1 Geographical Location

The former Aral Sea was located within the Turan Lowland, along the lower reaches of the Syrdarya river, which occupies its modern and ancient deltas (Fig. 2.1). Geographical coordinates are determined by the boundaries of the Kyzylorda region in Kazakhstan part: 42°40′–47°51′ N and 59°2′–68°9′ E. In the region, agricultural land occupied 12,996.5 thousand hectares, including irrigated arable land of 214.4 thousand hectares, of which 172.2 thousand hectares are engineered, perennial plantations of 2.9 thousand hectares, hayfields of 116.6 thousand hectares, and pastures of 12,580.6 thousand hectares. The leading crops in agriculture are the production of rice (68 thousand hectares) and vegetable and melon crops, and in animal husbandry, breeding, fine-wool sheep breeding, and beef cattle breeding (Faizov

G. Issanova et al., *Soil Cover of the Dried Aral Seabed in Kazakhstan*,
https://doi.org/10.1007/978-3-031-29867-7_2

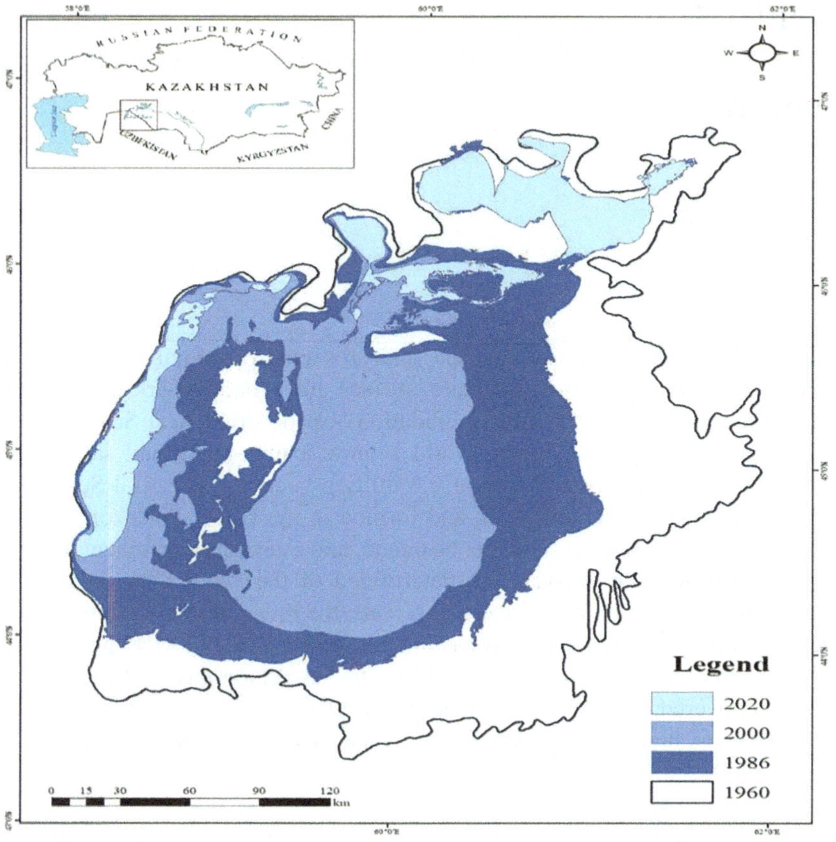

Fig. 2.1 The former Aral Sea and Aralkum desert

and Tapalova 2003). The region was divided into eight administrative regions with a predominantly agricultural economy with a developing oil and gas industry based on the Kumkol field (Faizov and Tapalova 2003).

The Aral Sea was located in the territory of Kazakhstan and Uzbekistan; before the lake began to recede, it was the fourth largest lake in the world. Since 1960s, the sea has started drying. In 1960, the water surface of the Aral Sea was approximately 69,000 km^2, while in 2020, the surface was an estimated 7000 km^2; only 10% of the seawater was left (Issanova et al. 2022). As a result, the Aral Sea almost dried up entirely due to the irrational use of water resources from the Syrdarya and Amudarya rivers for irrigation.

The dried Aral seabed is an ecologically critical area associated with a long period of the drying of the Aral Sea. It brings environmental problems particularly in Central Asian countries. Such anthropogenic changes in the hydrological regime of the Syrdarya and Amudarya rivers occurred. As a result, natural and man-made desertification has been noticed in the region. The Aral Sea region was observed as zones of intensive desertification and soil salinization leading to the growth of saline desert and deflation processes.

The processes of soil salinization and salt-dust removal of aerosols from the dried seabed, as dust-salt storms, environmental pollution with toxic chemicals, and destruction of the soil cover are progressing in the region.

2.2 Geological Structure, Rock Formations, and Soil-Forming Rocks

According to the geological structure, the Aral Sea region is located in the Syrdarya depression, which was formed in the Upper Paleozoic and subsequently filled with sediments of the Cretaceous, Paleogene, Neogene, and quaternary continental deposits. Rocks of the Paleozoic metamorphosed complex in the area of the Karatau mountain range, and tertiary sediments on the Cretaceous and Paleogene remnants, which are well expressed in the modern Syrdarya river delta, protrude from the pre-quaternary deposits to the surface. The latter are represented by red-colored gypsum clays (Cretaceous sediments) and greenish-gray Paleogene clays. The quaternary deposits, 3–10 m to 60–80 m thick, are distributed along the coast of the Aral Sea and are represented by heavily gypsum sands with interlayers of sandy loam, loam, and clay with shells of Cardium edule mollusks. They have also become widespread within the modern and ancient delta of the Syrdarya river (Faizov and Tapalova 2003).

Throughout the long geological history, most of the basin was an area of accumulation of terrigenous material, interspersed with the processes of erosional dissection and destruction (Borovsky and Pogrebinsky 1958). The latter, however, did not receive any significant depth here, and their areas did not disturb the general flat relief. As a result of tectonic movements and erosional calculation, delta-alluvial plains were formed, which subsequently determined the direction of lithomorphogenesis processes. In the Paleogene-Neogene time, an arid climate regime is established in the region. Moreover, as the Tien Shan Mountain system rose during the Alpine orogeny, the runoff of water from the mountain slopes led to the formation of the Syrdarya river, which filled the depression with alluvial deposits, with a thickness of 15–20 to 80–100 m. Alluvium has a two-membered structure: in the upper part, to a depth of 3–6 m, it consists of alternating interlayers of clays, loams, and silty (dusty) sands; coarse-grained sand and finer-grained sediments are in the lower part. The general pattern of alluvial sediments is the predominance of a light soil texture in areas adjacent to the channel and heavier (clayey) in interchannel depressions. The formation of the modern relief is closely related to the peculiarities of the geological structure of the territory, which in most of the region is characterized as a delta-alluvial plain.

The absolute heights of the plain vary from 200 m in the southeastern part to 53 m on the coast of the Aral Sea, with a surface slope in the northwest direction within 0.00022°. In the northern part, the territory is characterized as a

Fig. 2.2 Dried Aral seabed (Aralkum)

ridge-undulating table-remnant plain with absolute heights of 250–300 m. It is dissected by table hills, armored from the surface with sandstones, many dry depressions, valleys, and ravines (Faizov and Tapalova 2003).

Sandy massifs confined to gently sloping anticlines filled with Cretaceous, Paleogene, and Miocene sandy sediments are widespread (Aral Karakum, Small and Big Barsuks) (Fig. 2.2). In the western part, the Shagrai plateau stands out, strongly dissected by erosion and composed of Paleogene clays, limestones, and ferruginous sandstones. The entire eastern part of the depression is occupied by the Syrdarya delta-alluvial and Kyzylkum sandy plains. In connection with the continuing decrease in the level of the Aral Sea and the desertification of the territory, the land area is expanding, and a sandy-takyr-like solonchak landscape is being formed. The delta-alluvial plain is formed by the ancient (Daryalyk-Takyr and Bike-Sary plains) and modern (Kazaly) deltas and the valley of the Syrdarya river with takyr-like and takyr depressions scattered on their surface, dry channels, islands of sands, and knolls of Tertiary-Cretaceous remnants. The Syrdarya river valley has a wide floodplain and two terraces above the floodplain, dissected by a large number of channels, old rivers, lakes, riverbed ramparts, a ditch network, and small massifs of waving sands (Faizov and Tapalova 2003).

The Kyzylkum sandy desert is characterized by ridge-hummocky plain with dune relief in places, alternating with takyr depressions, ancient channels (Inkardarya, Kuvandarya, etc.), and islands of Tertiary-Cretaceous elevations. Remains of ancient fortresses and settlements and settlements of various historical epochs with traces of canals and irrigation fields have been preserved here (Faizov

1975). It is believed that the Kyzylkum sandy plain was formed as a result of the destruction and deflation of Paleogene bedrocks (Fedorovich 1983). In the formation of the relief here, an important role belongs to the accumulative activity of the Syrdarya river in the past and modern processes of deflation.

In the southeastern part of the Aral Sea region, the low-mountain ridge Karatau stands out with absolute heights from 400 to 600 to 1418 m. Their relative height does not exceed 300–400 m. folded Mesozoic sediments. The slopes of the ridge are cut by many narrow valleys of small rivers, which are lost at the foot of alluvial fans (Besaryk, Shelgaya, Zhidely).

The quaternary rocks formed the soil in the territory of the Aral Sea, mainly alluvial-deltaic, sandy, and layered sandy-argillaceous sediments, replaced on the remnants and slopes of the low mountains of Karatau by coarse skeletal deluvial and proluvial deposits. Their distinguishing feature is the predominance of dusty fractions (0.05–0.005 mm in size), a high degree of carbonate content (CO_2 up to 15–20%), and microaggregation in the composition.

Particles of physical clay (less than 0.01 mm in size) in soils and rocks are cemented into microaggregates due to carbonate content. At the same time, soil-forming rocks heavy (loamy, clayey) in soil texture consist of 80% water-resistant microaggregates; on lighter rocks (sandy and sandy loamy), they become easily pliable to wind erosion (Faizov and Tapalova 2003).

2.3 Surface and Groundwaters and Their Quality/ Quantity Changes

The surface waters of the territory under consideration form the drying Aral Sea, the lower reaches of the Syrdarya and Amudarya rivers. The water balance of the sea was determined by the runoff of the two main Central Asian rivers such as the Syrdarya (13.5 km^3) and the Amudarya (38.5 km^3). They formed the river flow of the Aral Sea basin. The rivers belong to the type of snow-glacial supply. The waters of these rivers in the Aral Sea basin irrigate over 50% of all irrigated lands in Central Asia, and water withdrawal for each hectare of the increase in irrigated area increased by 80,000 m^3 (Faizov and Tapalova 2003). The systematic increase in the area of irrigated land, the cultivation of moisture-intensive crops in the conditions of irrational use of water resources, caused the progressive drying of the Aral Sea and desertification of the territory. In addition, in the 1950s–1970s, a number of large water reservoirs (Kairakkum, Shardara, Charvak, Toktogul, Andijan, etc.) were built in the upper and middle reaches in the Syrdarya river, and Kazaly and Kyzylorda hydroelectric facilities were built. As a result, water withdrawal in the region has become comparable to the water resources of the basin (36.7 km^3), including non-returnable water consumption exceeding 20 km^3 of river water. Over the past quarter century, the influx of water to the ancient delta of the Syrdarya river decreased from 22–23 km^3 to 5–7 km^3, in the modern delta, respectively, from 15–16 km^3 to 1–3 km^3 (Popov et al. 1992).

The water area of the Aral Sea had an area of 66,000 km^2 in the 1950s–1960s, and the greatest length of 424 km and a width of 292 km, with an average depth of 16.1 m. Salinity of the seawater fluctuated within 10–11‰ (1%). In 1998, the salinity of the water in the northern half of the sea reached 65–67%. In the incoming balance of the sea, the runoff of river waters was 52 km^3, and atmospheric precipitation was 5.8 km^3. The expenditure part was estimated by evaporation values of 57.7 km^3 (Faizov and Tapalova 2003).

The degree of seawater mineralization in the period of 1981–1990 more than doubled compared with 1951–1960, and the chemical composition of water also changed in the direction of increasing toxic chlorine, magnesium, sodium, and phenol. The discharge of industrial, domestic, and collector-drainage waters into the river caused pollution with toxic salts, mineral fertilizers, herbicides, and defoliants. In the composition of the salts in the dried Aral seabed, sodium chloride-sulfate and magnesium chloride prevailed (Faizov and Tapalova 2003).

There are no systematic observations of the levels and salinity of groundwater at the bed of the Aral Sea. The latest observations are summarized by HydroInsGeo according to 1993–1996 data. They show that the falling sea level entails a decrease in the groundwater depression curve, keeping its values within 5 m from the surface. Higher groundwater levels are confined to the channels of the former watercourses and places of current releases from the lake complex toward the dried seabed. Mineralization of less than 10 g/L of groundwater takes place only along the routes of ancient channels and streams. On most of the dried seabed, mineralization exceeds 30 g/L (Dukhovny et al. 2008).

Intense salinization occurs due to evaporation from the groundwater table, where a groundwater level less than 3 m. The consequence is the development of solonetzes and solonchaks, common in numerous flat-bottomed depressions occupied by sors. The intensity of the manifestation of surface salinity processes in the Aral seabed reached 20–30% (Tomina 2009).

The sharp predominance of capillary rise in moisture determines the concentration of chlorine and sodium ions in the composition of groundwater, so they have a sulfate chloride type with high sodium content. The chemical composition of groundwater is generally identical to the seawater, despite the high content of individual ions. The tie-in into the bed of the riverbed led to the death of the previously operating channels and to a decrease in the groundwater level (the channels of the Kipchakdarya, Taldyk, Erkindarya, as well as the lower branches of Inzheneruzyak, Akkay). In recent years, the Amudarya flows along the straightened branch of the Urdabai. The influence of the inset is traced over a length of 100–20 km from the sea. According to the Temirbai gauging station, the bottom of the river has dropped to 3.8–4.5 m. Now the branches are fed by pumping stations. Simultaneously with the tie-in, there is a widespread decrease in the level of groundwater and pressure and flow rate of artesian wells in the Aral Sea region. The pressure water level, according to the forecast made in 1990, will decrease by 20–22 m 60–100 km from the sea coast (now 4–20 m), and the groundwater levels will drop to 3–5 m. Underground inflow into the sea basin is 0.2 km^3; however, due to the high concentration, 6.2 million tons of salts are carried out (Dukhovny et al. 2008). The thickness of the water-bearing stratum ranges from 1.0–3.0 m to 3.0–7.0 m. Groundwater in it occurs at depths from 3.0–4.0 m to 4.0–6.0 m and highly mineralized with a dry residue value from 17.0–34.0 g/L to 200 g/L. The filtration coefficient varies from 0.3–0.6 m/day to 0.8–3.0 m/day. The least mineralized waters with a salinity of 17.0–34.0 g/L are developed in Sudochie-Adjibai, Muynak, and Rybachinsky parts of the dried seabed (Dukhovny et al. 2008).

2.4 Climatic Conditions and Climate Change in the Region

The territory of the Aral Sea region is characterized by continental arid climate with hot, dry summers (high solar radiation and low humidity), cold winters, large intra-annual temperature fluctuations, little cloudiness, and negligible precipitation. The climatic conditions of the Aral Sea region are based on a variety of synoptic processes, rainfall patterns, and annual and interannual temperature conditions. The climate of the dried Aral seabed is very harsh. The seabed is located far from warm oceans that could affect its territory. The average annual temperature ranges from 5 to 11 °C, and the amount of precipitation is from 80–100 to 200 mm (Budyko 1974; Rachkovskaya 2003). Winter is long, from November to March without heavy snowfalls with cold north and northwestern winds. Hot, dry summers with scant precipitation and a large moisture deficit, frequent strong winds, and high solar insolation greatly increase evaporation. All these create sharply continental climatic conditions in the region (Small et al. 2001; He et al. 2022).

The wind regime is predominantly continental in nature. On average, the winds of the northeastern quarter dominate over the year with an average speed of 5–6 m/s. Strong winds often occur, and their frequency at Aktumsuk station reaches average 50 days/year, and at Barsakelmes, 44 days (Moloskova et al. 1987). Strong winds most often are observed in winter. In summer, breeze circulation exerts a noticeable influence on the microclimate of the coast, under the influence of which, in a zone up to 50 km wide. On the northern coast, the daytime breeze is southwesterly and the nighttime breeze is northwesterly. The daily change of wind directions at the stations of the southern coast is less abrupt. During the day, the wind blows from the northeast, and at night from the west. There is an increase in the number of dust storms with a maximum in April to July.

The desert climate is one of the leading factors in the formation and development of soil cover on exposed seabed surface as the lake played an important role in regulating the regional climate in the arid region (He et al. 2022). Since the change in lake surface area can affect local and regional near-surface air temperatures during the summer, air temperature in 2 m over the dried seabed was increased to >6 °C and to 1 °C in the 200 km around the original sea boundary toward to the south (Roy et al. 2014). The influence of the sea on the coastal territory in the northeastern and eastern parts extends over a distance of 80–100 km (Geldyeva et al. 1996). Significant changes in microclimate occurred in the process of retreat of the coastline by 40–60 km (Zhitomirskaya 1964; Ivanov et al. 1996; Moloskova and Il'nyek 1991;

Prokhorov 1972). The continuing shrinkage of the Aral Sea water area has greatly complicated the natural and climatic conditions of the coastal zone due to a decrease in overall moisture content, an increase in evaporation and removal of aerosol material, and a decrease in the level of groundwater.

The general climate change in the immediate vicinity of the former water level was reflected in an increase in the temperature of the summer months, especially March to August, and a decrease in temperature in November to December (Dukhovny et al. 2008). The transformation of the local climate in the dried seabed was accompanied by an increase in the frequency of sunny and very hot days by 15%, and the frequency of sunny wet days decreased by four times. The frequency of unfavorable weather for human activities has increased. However, the main patterns of atmospheric circulation associated with the movement of air masses in the north and south directions and conditions of moisture and temperature are also preserved. Fluctuations in the level of the Aral Sea, up to its complete drying, enhances the general continentality of the climate.

Based on our monitoring of climatic parameters, it can be confirmed that the climate of the northeastern dried Aral Sea bed is changing and getting harsh and more characteristic in arid zones and can be observed changes in weather conditions of the northeastern dried Aral Sea bed. Specifically for 34 years (from 1986 to 2020), temperature increased by 2 °C (from 8.4 to 10.4 °C), which also increased monthly mean potential evaporation by 76 kg m^{-2} (from 973 kg m^{-2} to 1049 kg m^{-2}), although monthly precipitation increased by 3 mm (from 10 to 13 mm) (Fig. 2.3). According to the most climate models, the temperature in arid Central Asia will increase by 1–2 °C by 2030–2050 (Lioubimtseva et al. 2005; Lioubimtseva 2014).

The monitoring of climatic parameters, specifically fluctuating the monthly mean precipitation (no stable trend) and increasing the monthly mean temperature and the monthly mean potential evaporation, has shown a process of aridization of the region. An increase in the average annual air temperature affects the intensification of aridization and degradation of arid ecosystems (Yas'kov 1999). The aridity of the climate in the region is high and, accordingly, the evaporation of groundwater is higher, so the likelihood of soil salinization is also high (Borovsky 1982).

2.5 Vegetation Cover and Its Degradation

The vegetation cover in the Aral Sea region is variable, dominated by semishrub communities, and is represented by saltwort-wormwood phytocenoses (xerophytic semishrub vegetation) with the participation of ephemers and ephemeroids. Due to the decreasing of the Aral Sea level and the increase in air dryness, the desert-sandy type of soil formation processes has become even more arid and saline.

Most of these shrubs are halophytes (including euhalophytes, hemihalophytes, and hyperhalophytes) and psammophytes, which grow in saline soils and sands, respectively. Over 20 species were identified within six plant communities in the dried Aral Sea bed, including perennial saltwort plants that are tolerant to saline

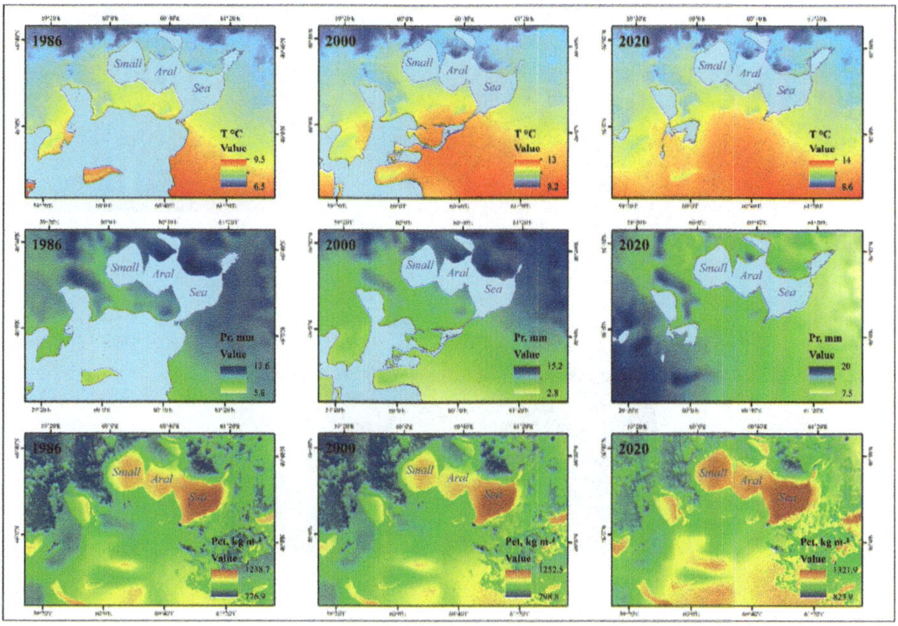

Fig. 2.3 Changes in climatic parameters: temperature, precipitation, and evapotranspiration in the northeast dried Aral Sea bed from 1986 to 2020

soils (e.g., Anabasis salsa, Halostachys belangeriana Botsch., Halocnemum strobilaceum Pall., Climacoptera lanata Botsch., Climacoptera brachiate Botsch., and Suaeda microphylla Pall.) and mugwort plants (Artemisia terrae-albae Krasch., Ceratocarpus arenarius L., Kochia prostrata L., Allium sabulosum Bunge, Ceratocarpus utriculosus Bluket, Aeluropus littoralis Parl., Alhagi pseudalhagi Fisch., Petrosimonia brachiata Bunge, and Salsola nitraria Pall.), including psammophytic plants (Artemisia terrae-albae Krasch., Petrosimonia brachiata Bunge, Salsola nitraria Pall., and Kochia prostrata L.), black saxauls (Haloxylon aphyllum Minkw., Peganum harmala L., Salsola paulsenii Litv., and Alhagi pseudalhagi Fisch.), mixed saxauls (Haloxylon aphyllum, Halostachys belangeriana Botsch., and Salsola paulsenii Litv.), and psammophytic shrub plants (Tamarix hispida Willd., Zygophyllum fabago L., Girgensohnia oppositifora Pall., Nitraria sibirica Pall., Petrosimonia sibirica Pall., and Astragalus orbiculatus Ledeb.) (Fig. 2.4). However, a significant portion of the study area is not occupied by plants at all. According to the study results, all these plant communities can be found in predominantly sulfurous soils, due to their less harmful effects on vegetation cover. Perennial saltworts, psammophytic shrubs, and black saxaul communities occur in predominantly chloridic soils, while the last two groups also occur in sands. However, perennial saltwort communities are found only in soda soils, confirming their hyperhalophytic and eugalophytic capacities.

Haloxylon aphyllum (Minkw.) *Halocnemum strobilaceum* *Calligonum aphyllum*

Tamarix hispida

Fig. 2.4 Dominant plant species in the Aral Sea region in Kazakhstan

According to the disturbed vegetation cover assessment results, the degree of vegetation degradation can be divided into four classes: slight, temperate, significant, and strong degradation. The corresponding percentages of the disturbed vegetation cover were as follows: the slight class comprised 0–3% of the area, the temperate class included 3–10%, the significant degradation class corresponded to 10–20% of the area, and strong vegetation degradation was identified in >20% of the area (NAKZ 2010). The study results and comprehensive analysis showed that 21.2% and 7% of the analyzed vegetation cover had degraded to the significant and strong degrees, respectively.

Perennial saltwort plants dominated in the study area, and most of the dried seabed was covered with these plants, corresponding to the mostly degraded class; 14.3% of the area had undergone strong degradation, while 12.4% of the area had undergone significant degradation. Of the saxaul (black and mixed) vegetation cover, 19.4% and 16% corresponded to the strong and significant degradation degrees, respectively. In addition, the majority of psammophytes shrub plants had degraded to a significant degree (69%) (Fig. 2.5).

In the northern part of the region, ephemeroid-sagebrush groups are widespread on brown desert soils. In combination with them, biyurgun and black wormwood associations are found on takyr-like plains; on Tertiary-Cretaceous remnants, they are replaced by biyurgun-wormwood-boylysh communities on gray-brown soils.

The vegetation cover of sandy massifs within the subzone of brown soils (Aral Karakum, Borsuk) is formed by erkekovo-wormwood-sandy acacia, erkekovo-wormwood-zhuzgun, and erkekovo-wormwood-shrub (bindweed, kurchavka,

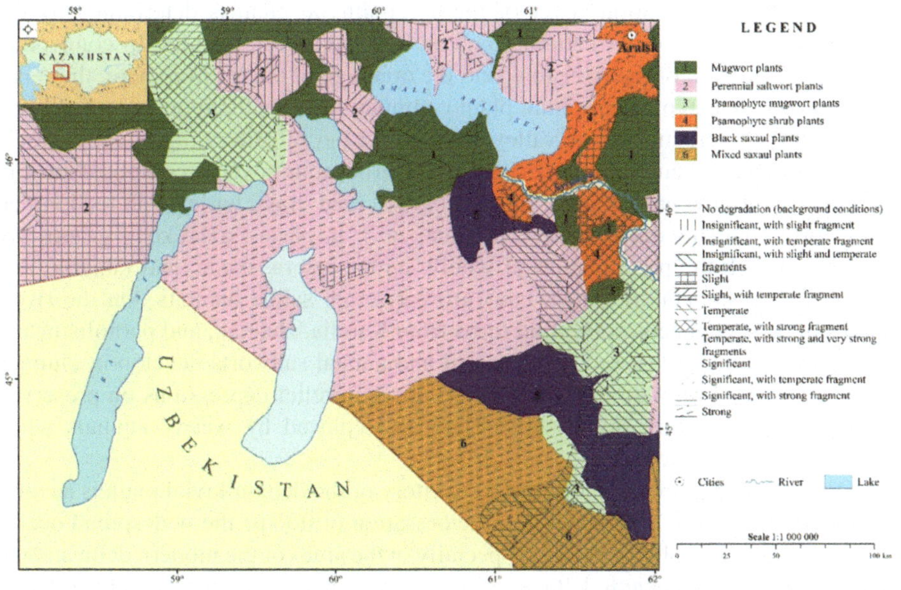

Fig. 2.5 Plant communities and their degradation (Issanova et al. 2022)

zhuzgun) associations, with a productivity of 8.7 c/ha. Large areas are occupied by erkek communities with a yield of 3–8 q/ha. Hilly and hilly-ridged sands of the subzone of gray-brown soils characterize ephemeral-sagebrush-shrub communities.

Erkek falls completely here, large areas are occupied by saxaul light forests (black in the northern part, white saxaul in the south), zhuzgun and boyalych, with a well-developed ephemera cover of bulbous bluegrass. On semifixed sands, ephedra, eremurus, and Zhongarskaya wormwood are found (Kurochkina 1978), and on saline habitats, azhrek, zhantak, and juicy saltwort. Among the takyr-like plains of the northern Kyzylkum, along the dried-up channel of the Zhanadarya, in a strip 250 km wide, on an area of about 300,000 ha, there was previously a unique, largest in Central Asia, now a drying massif of black saxaul forest. Its existence was supported by periodic flooding of the Zhanadarya channel, especially during the high floods of the Syrdarya river. Fresh groundwater filtered through the bed and sides of the channel and replenished groundwater reserves, maintaining their level at a depth of 7–15 m, which was the most favorable for the development of saxaul. At present, no water flows into the channel of the Zhanadarya, groundwater has decreased, and the black saxaul forest is dying. According to the data on the composition of saxaul, ash, chlorides, and alkalis predominate; in biyurgun litter, chlorides and sulfates are contained in almost equal amounts with an increased value of sodium and potassium alkalis. Tamarix plants are distinguished by an almost 30fold predominance of alkaline earth bases over alkaline and 3.3 times chlorides over sulfates. The chemical composition of plant ash significantly affects the genetic properties of soils.

The modern vegetation cover of the delta-alluvial plain is determined by the simultaneous influence of natural zonal and anthropogenic factors. In connection with the regulation of the river flow and the drainage of the territory here, there is a widespread change of hydrophilic plant communities to halophytic and xerophytic ones. On the riverbanks of the valley in the lower tier of gallery tugai forests (loch, willow, turanga, zhengil, tamariks, etc.), the herbaceous cover is composed mainly of grasses with the participation of couch grass, reed grass, barley, and reed mixed with forbs (marshmallow, kendyr, licorice, etc.). In places, azhrekovy groups are developed with the participation of woolly climacoptera, sveda, and tamarix, and halophytes (karabarak, kermek, etc.) grow in more saline habitats. On the riverbanks of dead channels, xerophilic forbs from Karelia, zhantak, and parnolistny are widespread, and in deserted areas, jengil and annual saltworts develop in clumps. Reed and cattail formations form a dense cover in relief depressions on preserved bog and meadow-bog saline soils. They are replaced by weedy annuals when drained and desertified.

Drainage and aridization of the vast territory of the delta-alluvial plain is accompanied by the death of channels and the cessation of floods; the widespread occurrence of solonchak phytocenoses, especially in the areas of the modern delta and the western part of the ancient delta; and a decrease in land productivity and soil fertility. With natural moistening, reed meadows here annually synthesized up to 125 c/ha of aboveground and 250 underground organic matter, which maintained the high fertility of hydromorphic soils. At the same time, in the drained areas, the productivity of weedy weed grass does not exceed 50 centners per hectare in mass, including 13 aboveground and 37 underground. Moreover, the chemical composition of plant litter mineralized in the soil contains a lot of sodium, which forms toxic salts with chlorine ions and soil carbonates, which contribute to solonetzation and soil salinization.

At present, as a result of many years of irrational economic use, the vegetation cover has turned out to be strongly transformed. Full-membered climax plant communities have been preserved in fragments; various anthropogenic modifications predominate in the vegetation cover. Despite the floristic and structural difference between the initial types of vegetation and the initial stages of their digression, communities eventually converge, accompanied by a loss of floristic diversity and a simplification of their composition and structure. In large areas, the absolute dominance of weed species (adraspan, ebelek, itsigek, etc.) is noted. In this regard, the vegetation cover loses its resource forage and indicator significance, and the gene-coenofund loses its functional role in the landscape and biosphere. In the Aral Sea region, desertification becomes catastrophic, covering over 60% of the territory. A very strong degree of desertification is noted in the valley and delta of the Syrdarya river where there is a change of meadow-tugai type of vegetation to desert. Reed, forb-grass, licorice meadows, and tugai thickets have almost disappeared here, which are almost impossible to restore. Reducing the biological productivity and area of tugai forests leads to a decrease in river flow and water deficit in the region.

A strong degree of desertification is accompanied by the formation of unstable communities, in which weed and poisonous species are abundant. Areas with a moderate degree of desertification occur in the central and southwestern Kyzylkum and western and partly southern coasts of the Aral Sea. The vegetation cover here is represented by long-term derived, more or less stable communities. A weak degree of desertification is characterized by the presence of slightly modified plant groups in hard-to-reach and waterless areas.

References

Borovsky VM (1982) Formation of saline soils and halogeochemical provinces of Kazakhstan. Science of KazSSR, Almaty, p 253. (in Russian)

Borovsky VM, Pogrebinsky MA (1958) The ancient delta of the Syr Darya and the Northern Kyzyl Kum. Alma-Ata. Academy of Sciences KazSSR, p. 72–77 (in Russian)

Budyko MI (1974) Change of climate. Hydrometeoizdat, Leningrad. in Russian

Dukhovny VA, Navratil P et al (eds) (2008) Comprehensive remote sensing and ground based studies of the dried Aral Sea bed. Scientific-Information Center ICWC, Tashkent

Faizov KS (1975) About soils of ancient irrigation lands. J Bull Acad Sci KazSSR 1:39–45. (in Russian)

Faizov KS, Tapalova AS (2003) The ecological crisis in Aral Sea region and their solution problems. Almaty, p 110. (in Russian)

Fedorovich BA (1983) Dynamics and patterns of relief formation of deserts. Nauka, Moscow, p 238. (in Russian)

Geldyeva GV, Budnikova TI, Medvedeva NY (1996) The main regularities of natural stabilization of the landscapes of the dried Aral seabed. J Vestnik KazGU Ser Geogr 3:34–40

He H, Hamdi R, Luo G, Cai P, Zhang M, Shi H, Li C, Termonia P, De Maeyer P, Kurban A (2022) Numerical study on the climatic effect of the Aral Sea. Atmos Res 268:105977. https://doi.org/10.1016/j.atmosres.2021.105977

Issanova G, Abuduwaili J, Tynybayeva K, Kalybayeva A, Saduakhas A, Kulymbet K, Kaldybayev A, Erlan G, Tanirbergenov S (2022) Soil salinisation as a land degradation process in the dried bed of the North-eastern Aral Sea, Kazakhstan. Arab J Geosci 15:1055. https://doi.org/10.1007/s12517-022-09627-w

Ivanov V, Chub V et al (1996) Review of the scientific and environmental issues of the Aral Sea basin. In: The Aral Sea basin, NATO ASI series, vol 2. Springer, Berlin, pp 9–21

Kurochkina LY (1978) Psammophilic vegetation of Kazakhstan. Alma-Ata: Nauka, 272 p (In Russian)

Lioubimtseva E (2014) Impact of climate change on the Aral Sea and its basin. In: Micklin P et al (eds) The Aral Sea, Springer earth system sciences. Springer, Berlin Heidelberg. https://doi.org/10.1007/978-3-642-02356-9_17

Lioubimtseva E, Cole R, Adams JM, Kapustin G (2005) Impacts of climate and land-cover changes in arid lands of Central Asia. J Arid Environ 62(2):285–308. https://doi.org/10.1016/j.jaridenv.2004.11.005

Moloskova TI, Il'nyek EP (1991) Climatic fluctuations in the general circulation of the atmosphere and types of synoptic processes in Central Asia. Works of SARNIIHMI, Hydrometizdat, Tashkent, No. 141, pp 3–102. (in Russian)

Moloskova TI et al (1987) Climatic consequences in the Aral Sea region. Gidrometeoizdat, Moscow, p 119. (in Russian)

National Atlas of Kazakhstan (NAKZ) (2010) Environment and ecology, vol 3. NAKZ, Almaty. in Russian

Popov YM, Nekrasova TF, Semenov OE, Starodubtsev VM (1992) Anthropogenic changes in the soils of the Aral Sea region and their ecological and economic significance, Almaty, p 60. (in Russian)

Prokhorov II (1972) Possible changes in individual climatic characteristics due to changes in the area of the Aral Sea. J Questions of climate fluctuations and water resources, L.: Gidrometeoizdat, pp 44–49. (in Russian)
Rachkovskaya EI (2003) Botanical – geographic types of deserts. In: Rachkovskaya EI, Volkov EA, Khramtsov VN (eds) Botanical geography of Kazakhstan and Middle Asia (desert region). Springer, Berlin, pp 26–28
Roy SB, Smith M, Morris M, Orlovsky N, Khalilov A (2014) Impact of the desiccation of the Aral Sea on summertime surface air temperatures. J Arid Environ 110:79–85. https://doi.org/10.1016/j.jaridenv.2014.06.008
Small E, Giorgi F, Sloan LC, Hostetler S (2001) The effects of desiccation and climatic change on the hydrology of the Aral Sea. J Clim 14:300–322
Tomina TK (2009) Soils of the dried Aral seabed. J Hydrometeorol Ecol 1:60–75. (in Russian)
Yas'kov MI (1999) Desertification of Shu Valley (Mountain Altay), Russia, Biysk, p 195. (in Russian)
Zhitomirskaya OM (1964) Climatic description of the Aral Sea region. L.eningrad: Hydrometeoizdat, p 68. (in Russian)

Chapter 3
Origin of the Aralkum Desert (Dried Aral Seabed)

Abstract Since the 1960s, the Aral Sea has been shrinking due to irrational irrigation. In 1989, the sea was separated into Big and Small Aral seas. From 1960 to 2020, the seawater surface area has shrunk by 10 times. Now only Big Aral Sea has left and rest part of the former sea is a salty desert – Aralkum. The desert is considered as a potential source of storms with salt, dust, and sand particles. The highest amount of dust transport was observed at Uialy, Zhusaly, and Kyzylorda weather stations, from 1143 tons km^{-1} $year^{-1}$ to 2684 tons km^{-1} $year^{-1}$, respectively.

Keywords Aral Sea · Aralkum · Shrinkage · Water salinity · Dust and salt storms

3.1 Shrinkage of the Aral Sea

Internal lakes are an important part of water resources in arid or semiarid areas (Bai et al. 2011; Zhan et al. 2021; Yue et al. 2021). In recent years, due to climate warming and irrational development and human use, several internal lakes are on the verge of drying up (Moloskova and Il'nyek 1991; Small et al. 2001; Bosch et al. 2007; Lioubimtseva et al. 2005; Lioubimtseva 2014; He et al. 2022; Aladin et al. 2019). One of them is the Aral Sea, which was one of the largest internal lakes in the world in the past (Ivanov et al. 1996; Micklin 2010). In 1960, the recorded maximum level of the lake was 53 m and the surface area was about 67,000 to 68,000 km^2, depending on the data source and the methodology used to estimate its area, and the salt content was 1% (Micklin 2004, 2010; Cretaux et al. 2013). Between 1960 and 2004, the water surface area decreased by 8.5 times, and the water level fell by approximately 20 m as water from upstream rivers began to be diverted for irrigation. In 1989, the Aral Sea split into two isolated reservoirs: the North (Small) and South (Big) Aral Seas (Micklin 2010; Wu et al. 2021).

Kokaral is the first of the large islands, having connected in the west with the coastline, which became a peninsula isolating the Small Sea in the west. By 1986, the peninsula and in the east had practically separated from the Big Aral Sea and only a narrow stream remained. Starting from this time, the hydrological regime of the Small

G. Issanova et al., *Soil Cover of the Dried Aral Seabed in Kazakhstan*,
https://doi.org/10.1007/978-3-031-29867-7_3

and Big Seas differs markedly. The construction of the Kokaral dam, 12 km long and 8 m high, which separated the Small and Big Seas, changed the hydrological regime of water bodies. Despite repeated breaks of the dam, the level of the Small Aral Sea in 1990–2000 was somewhat higher than in the Big Aral Sea. The area of the Small Sea in these years varied within 2900–3200 km^2 (Chub 2000; Roy et al. 2014).

Since the 1960s, the expansion of irrigated agriculture has led to huge water diversions from the Aral Sea's only two inflows, a major causal factor in its decline (Micklin 2008). Between 1960 and 2009, the shrinking Aral Sea exposed almost 60,000 km^2 of its former seabed of 67,499 km^2 (Micklin 2010), which is now a highly unstable landscape with a high degree of ecological hazard in terms of desertification and aeolian erosion (Dukhovny et al. 2008) (Issanova et al. 2020).

A huge new solonchak desert "Aralkum" has emerged on the former seabed, with an area estimated at more than 57,000 km^2 in 2011 (Indoitu et al. 2015; Breckle et al. 2001), and is considered the largest area of primary vegetation change in the world (Breckle et al. 2012). These newly formed landscapes are considered a major source of salt and dust storms (Semenov 2012; Singer et al. 2003; Issanova et al. 2015). Frequent releases of salt/sand dust and aerosols pose a serious threat to the ecological environment and local residents in the Aral Sea region (Löw et al. 2013; O'Hara et al. 2000; Wiggs et al. 2003). The vulnerability of the Aral Sea region and the local human society has now increased to a critical level. The mitigation of this anthropogenic crisis requires timely monitoring information and effective management measures.

Remote sensing technology has provided a way for large-scale geospatial analysis and dynamic environmental monitoring of the Aral Sea surface area (Berdimbetov et al. 2021). In the early 1970s, satellite imagery was successfully used to monitor water regression and desertification in the Aral Sea basin (Shi and Wang 2015; Shi et al. 2014).

The main morphometric data, water level, water area and volume, and water salinity of the Aral Sea, for comparative analysis were taken from the website of the Regional Information System on Water and Land Use in the Aral Sea Basin (CAWATERinfo). Based on these data, the changes in water parameters were analyzed over the following periods: 1960–1968, 1969–1987, and 1988–2010.

In analyzing the water data from 1960 to 2010, the degradation of water parameters is particularly evident. On the basis of the water area data from 1960 to 2010, a downward trend in water is generally noticeable. However, according to the trend line, the period can be divided into three intermediate periods, where from 1960 to 1968, there is a steady decline in area. From 1969 to 1987, the trend improves relatively, and from 1988 to 2010, the decrease in area is reversed (Fig. 3.1a). A similar process is observed in water depth. Intermediate periods (1960–1968, 1969–1987, and 1988–2009), where decrease and relative increase/stable condition of water level and water area were observed, are the same (Fig. 3.1b). The decline in water volume can be divided into four intermediate periods. From 1960 to 1968, the water level gradually decreased, while from 1969 to 1978, the decreasing trend was relatively stable and even an increasing one can be observed in some places. From 1979 to 2000, the process reverses a negative trend and it is only since 2001 that the volume of water in the Aral Sea has been regulated (Fig. 3.1c). This is closely linked to

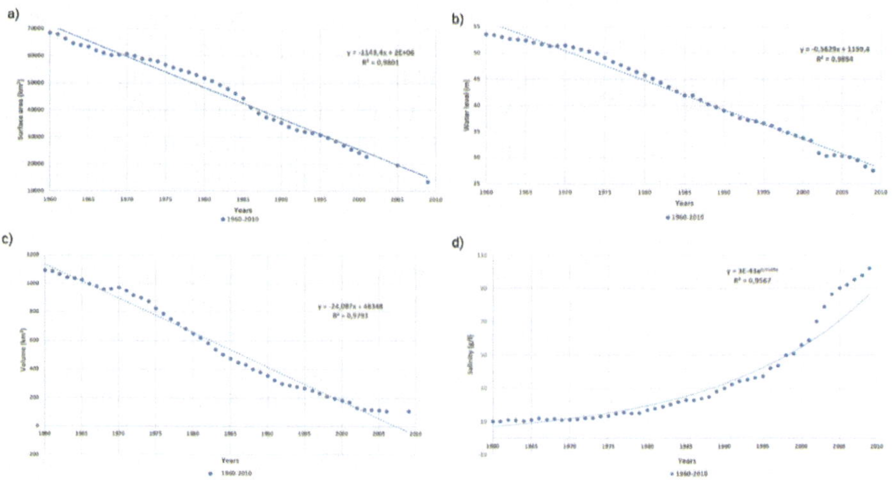

Fig. 3.1 Basic morphometric parameters of the Aral Sea for 1960–2010

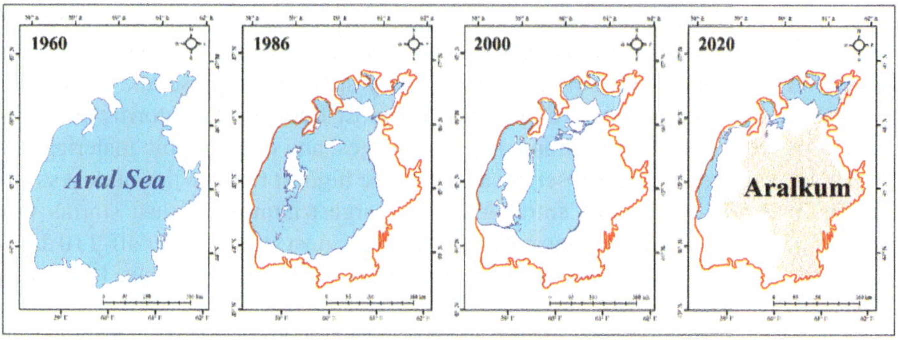

Fig. 3.2 Changes in water surface area for different periods

the construction of the Kokaral dam in 2005. The salinity of the water has increased almost every year since 1960 and by 2009 reached 102 g/L (Fig. 3.1d).

NDWI values were calculated from Landsat data for three different periods, 1986, 2000, and 2020, to determine and map the water surface area of the Aral Sea (Fig. 3.2). The data obtained showed that the water surface area of the Aral Sea was shrinking at a rapid rate, and comparing the data for 1960 and 2020, we see that the water surface area has shrunk by 10 times.

3.2 Aralkum as a Source of Dust and Salt Storms

Dust/sand and salt storms are common events in the arid regions and sandy and solonchak deserts, including Aralkum desert. The dried Aral seabed (Aralkum) is a huge open salt surface and consists mainly of solonchak desert (70%) and sand/sand–loamy deserts (20%) (Micklin 2010). This solonchak desert is the source of salt and dust storms and deposits affecting all of the adjacent regions, including the quality of the environment (climate and landscapes, water, soil and air quality, quality of agriculture and livestock products) and consequently the living conditions, health, and economic activities of the population in the region. Among soil/land degradation, soil salinization and wind erosion with the anthropogenic genesis is dominant in the region (Medeu 2010; Breckle et al. 2012; Issanova et al. 2022).

The environmental conditions (high wind speed regime, scarcity of vegetation cover, frequency of soil, and atmospheric droughts) in the desert promote the development of storms. These natural and anthropogenic factors determine the high vulnerability of the territory to the development of deflation processes as dust and sand storms.

The size and amount of blown particles during storms depends mainly on the soil-surface structure and wind regimes. Strong winds are often recorded in the region with mean wind speeds reaching from 6 to 7 m s^{-1} at weather stations in the region. Maximum wind speeds can reach 20–25 m s^{-1} in the summer (Galayeva et al. 1996; Galayeva and Semenov 2011).

In general, the annual number of days with dust storms is 55–60 in the desert zone (near the Aral Sea and Balkhash Lake regions) (NAKZ 2010). These areas are with most frequent dust storms, where wind speed is high and soils have light texture, which are under intensive use, or in sandy areas with sparse and halophytic vegetation since the areas were identified for most frequent dust storms based on data of meteorological observation, satellite images, and cartographic materials.

The Aralkum desert was observed as one of the highest frequencies of dust storm regions in Kazakhstan and Central Asia. The largest number of dust storms was observed in the Aral Sea weather station (WS) where dust storms occur 40–110 days/year (Fig. 3.3). They can happen in summer when wind speed reaches 10–14 m/s (Babaev 1999).

According to the analyzing results, the direction and moving of the sand/dust/salt can determine the impact of storms in the surrounding territories. The resultant vector also shows the final direction of moving aeolian landforms. The sand/dust/salt transportation prevails in the west and southwest directions in almost weather stations, except Aral Sea WS (Fig. 3.4). This may be due to specific synoptic processes in the region.

The Aral Sea and Uialy WS are considered as coastal stations; the rest of the stations (Zhusaly, Kyzylorda, Kazaly, Shirik-Rabat, and Karak) are considered continental stations. The Karak and Shirik-Rabat WSs are located in a region where the anthropogenic influence on the generation of dust is minimal and there is a little impact of removal from dry areas of the Aral Sea (Galayeva and Semenov 2011). The Kyzylorda WS is analogous to the Karak MS, but the amount of sand transport is somewhat greater and the maximum sand transport in the west and southwest direction was 629 tons km^{-1} $year^{-1}$ (Galayeva and Semenov 2011).

On the southernmost Shirik-Rabat WS, the sand transport is dominant in a southwesterly direction with 377 tons km^{-1} $year^{-1}$, and in rest of directions, the transfer is very small. A western direction was dominant for Zhusaly WS in the amount of

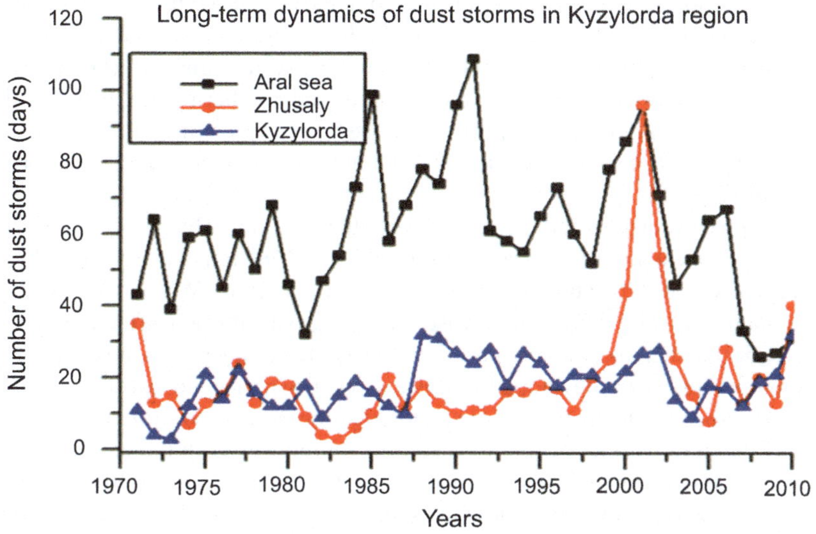

Fig. 3.3 Long-term dynamics of dust storms in the Aral Sea region (Issanova et al. 2015)

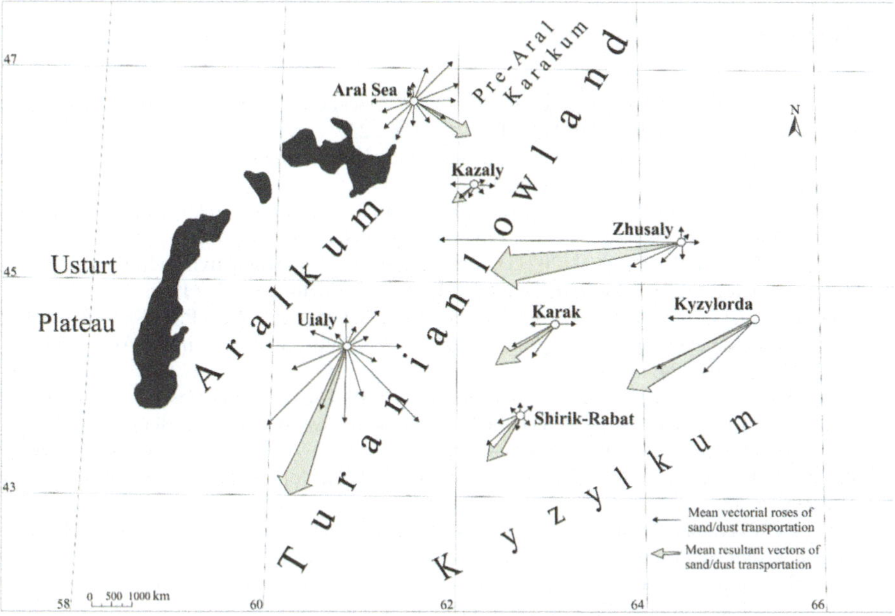

Fig. 3.4 Wind directions in the dried Aral seabed (Aralkum) (Issanova et al. 2017)

2684 tons km^{-1} year^{-1}. Because fixed sands surround the Kazaly WS, it had the lowest maximum at 86 tons km^{-1} year^{-1} in west and southwest directions. The southeast (1143 tons km^{-1} year^{-1}) and southwest (967 tons km^{-1} year^{-1}) directions were dominant at the Uialy WS.

A large proportion of the sand transported in the southeast direction was due to a very intense dust storm on June 15–17, 1984. It then was transferred at 29,000 tons km^{-1} $year^{-1}$, of which 57% is the average long-term transport (Galayeva and Semenov 2011). In general, in the most central and southern parts of the Aral Sea region, the high amount of transport was in the west and southwest directions; only a few weather stations had a maximum amount in the southeast, east, and northeast directions (Fig. 3.4).

The highest transport intensity was exhibited at Uialy, Zhusaly, and Kyzylorda WSs. The greatest amount of transported sand was 2684 tons km^{-1} $year^{-1}$ observed at Zhusaly MS at 270° and 1143 tons km^{-1} $year^{-1}$ at Uialy WS at 135° (Galayeva and Semenov 2011). These new, vast, sandy massifs formed after the sea level decreased, thus leaving them exposed and prone to weathering. Consequently, the deflation processes are developed in this region. Thus, the study of dust- and sand-transport directions leads to the conclusion that the process of moving sands in the north of the 46th parallel occurs in east direction; however, in the eastern Aral Sea region, the movement of sand, which is located in the south of the parallel, is observed in the west direction with the exception of Uialy WS where transfer occurs to the south.

References

Aladin NV, Gontar VI, Zhakova LV et al (2019) The zoocenosis of the Aral Sea: six decades of fast-paced change. Environ Sci Pollut Res 26:2228–2237

Babaev AG (ed) (1999) Desert problems and desertification in Central Asia: the researches of the desert institute. Springer, Berlin/Heidelberg

Bai J, Chen X, Li JL, Yang L, Fang H (2011) Changes in the area of inland lakes in arid regions of Central Asia during the past 30 years. Environ Monit Assess 178(1–4):247–256

Berdimbetov T, Ilyas S, Ma Z et al (2021) Climatic change and human activities link to vegetation dynamics in the Aral Sea basin using NDVI. Earth Syst Environ 5:303–318

Bosch K, Erdinger L, Ingel F, Khussainova S, Utegenova E, Bresgen N, Eckl PM (2007) Evaluation of the toxicological properties of ground- and surface-water samples from the Aral Sea basin. Sci Total Environ 374:43–50

Breckle SW, Wucherer W, Agachanjanz O, Geldyer B (2001) The Aral Sea crisis region. In: Breckle SW, Veste M, Wucherer W (eds) Sustainable land use in deserts. Springer, Berlin, pp 27–37

Breckle SW, Wucherer W, Dimeyeva LA, Ogar NP (2012) Aralkum – a man-made desert: the desiccated floor of the Aral Sea (Central Asia), Ecological studies. Springer, Berlin, Heidelberg, 488 p

Chub VE (2000) Climate Change and its Impact on the Natural Resources Potential of the Republic of Uzbekistan. Hydrometeorological Service of Uzbekistan, 219 p

Cretaux J-F, Letolle R, Bergé-Nguyen M (2013) History of Aral Sea level variability and current scientific debates. Global Planet Change 110(Part A):99–113

Dukhovny VA, Navratil P et al (eds) (2008) Comprehensive remote sensing and ground-based studies of the dried Aral Sea bed. Scientific-Information Center ICWC, Tashkent

Galayeva OS, Semenov OE (2011) Sand transportation by wind in the Aral Sea region. Hydrometeorol Ecol 3:73–85. (in Russian)

Galayeva OS, Semenov OE, Shapov AP (1996) Features of sand transportation in the Aral Sea region. Hydrometeorol Ecol 4:73–93. (in Russian)

He H, Hamdi R, Luo G, Cai P, Zhang M, Shi H, Li C, Termonia P, De Maeyer P, Kurban A (2022) Numerical study on the climatic effect of the Aral Sea. Atmos Res 268:105977. https://doi.org/10.1016/j.atmosres.2021.105977

Indoitu R, Kozhoridze G, Batyrbaeva M, Vitkovskaya I, Orlovsky N, Blumberg D, Orlovsky L (2015) Dust emission and environmental changes in the dried bottom of the Aral Sea. Aeolian Res 17:101–115

Issanova G, Abuduwaili J et al (2015) Aeolian transportation of sand and dust in the Aral Sea region. Int J Environ Sci Technol 12(10):3213–3224

Issanova GT, Abuduwaili J, Mamutov ZhU, Kaldybaev AA, Saparov GA, Bazarbaeva TA (2017) Saline soils and identification of salt accumulation provinces in Kazakhstan. Arid Ecosyst 7(4):243–250

Issanova G, Saduakhas A et al (2020) Desertification and land degradation in Kazakhstan. J Bull Natl Acad Sci Republic of Kazakhstan 5(387):95–102

Issanova G, Abuduwaili J, Tynybayeva K, Kalybayeva A, Saduakhas A, Kulymbet K, Kaldybayev A, Erlan G, Tanirbergenov S (2022) Soil salinisation as a land degradation process in the dried bed of the north-eastern Aral Sea, Kazakhstan. Arab J Geosci 15:1055. https://doi.org/10.1007/s12517-022-09627-w

Ivanov V, Chub V et al (1996) Review of the scientific and environmental issues of the Aral Sea basin. In: The Aral Sea basin, NATO ASI series, vol 2, pp 9–21

Lioubimtseva E (2014) Impact of climate change on the Aral Sea and its basin. In: Micklin P, Aladin N, Plotnikov I (eds) The Aral Sea, Springer earth system sciences, vol 10178. Springer, Berlin

Lioubimtseva E, Cole R, Adams JM, Kapustin G (2005) Impacts of climate and land-cover changes in arid lands of Central Asia. J Arid Environ 62(2):285–308. https://doi.org/10.1016/j.jaridenv.2004.11.005

Löw F, Navratil P, Kotte K, Schöler HFF, Bubenzer O (2013) Remote-sensing-based analysis of landscape change in the desiccated seabed of the Aral Sea – a potential tool for assessing the Hazard degree of dust and salt storms. Environ Monit Assess 185(10):8303–8319

Medeu AR (ed) (2010) Republic of Kazakhstan, vol 3. Environment and Ecology, Almaty, p 520. (in Russian)

Micklin P (2004) The Aral Sea crisis. In: Nihoul CJ, Zavialov P, Micklin P (eds) Dying and dead seas: climatic vs. anthropic causes, NATO science series IV: earth and environmental sciences, vol 36. Kluwer, Boston, pp 99–125

Micklin PP (2008) Using satellite remote sensing to study and monitor the Aral Sea and adjacent zone. In: Qi J, Evered KT (eds) Environmental problems of Central Asia and their economical, social and security impacts. Springer, Heidelberg, pp 31–49

Micklin P (2010) "The past, present, and future Aral Sea", lakes & reservoirs: science. Policy Manage Sustain Use 15(3):193–213

Moloskova TI, Il'nyek EP (1991) Climatic fluctuations in the general circulation of the atmosphere and types of synoptic processes in Central Asia. Works of SARNIIHMI, Gidrometizdat, No. 141, pp 3–102. (in Russian)

National Atlas of Kazakhstan (NAKZ). Environment and ecology. 2010. Almaty, Volume 3. (in Russian)

O'Hara SL, Wiggs GFS, Mamedov B, Davidson G, Hubbard RB (2000) Exposure to airborne dust contaminated with pesticide in the Aral Sea region. Lancet 355:627–628

Roy SB, Smith M, Morris M, Orlovsky N, Khalilov A (2014) Impact of the desiccation of the Aral Sea on summertime surface air temperatures. J Arid Environ 110:79–85. https://doi.org/10.1016/j.jaridenv.2014.06.008

Semenov OE (2012) Dust storms and sandstorms and aerosol long-distance transport. In: Breckle SW, Wucherer W, Dimeyeva LA, Ogar NP (eds) Aralkum – a man-made desert: the desiccated floor of the Aral Sea (Central Asia), Ecological studies, vol 218. Springer, Berlin, pp 73–82

Shi W, Wang M (2015) Decadal changes of water properties in the Aral Sea observed by MODIS-Aqua. J Geophys Res Oceans 120:4687–4708

Shi W, Wang MH, Guo W (2014) Long-term hydrological changes of the Aral Sea observed by satellites. J Geophys Res Oceans 119(6):3313–3326
Singer A, Zobeck T, Poberrezky L, Argaman E (2003) The PM10 and PM2.5 dust generation potential of soils/sediments in the southern Aral Sea basin, Uzbekistan. J Arid Environ 54:705–728
Small E, Giorgi F, Sloan LC, Hostetler S (2001) The effects of desiccation and climatic change on the hydrology of the Aral Sea. J Clim 14:300–322
Wiggs GFS, O'Hara S, Wegerdt J et al (2003) The dynamics and characteristics of aeolian dust in dryland Central Asia: possible impacts on human exposure and respiratory health in the Aral Sea basin. Geogr J 169:142–157
Wu Q, Yue H, Liu Y, Hou E (2021) Geospatial quantitative analysis of the Aral Sea Shoreline changes using RS and GIS techniques. Earth Sci Inf. https://doi.org/10.1007/s12145-021-00714-2
Yue H, Liu Y, Wei J (2021) Dynamic change and spatial analysis of great lakes in China Based on Hydroweb and Landsat Data. Arab J Geosci 14:149
Zhan S, Wu J, Jin M (2021) Hydrochemical characteristics, trace element sources, and health risk assessment of surface waters in the Amudarya Basin of Uzbekistan, arid Central Asia. Environ Sci Pollut Res, vol 29, pp 5269–5281

Chapter 4
Soil Cover of the Dried Aral Seabed (Aralkum)

Abstract The dried bed of the Aral Sea is located mainly in flat and undulating ridged plains. The main soil type is solonchaks with different types, and sands, and meadow-floodplain-meadow soils. Soil environment is alkaline (7.3–8.2). The absorption capacity is saturated with calcium, magnesium, and sodium. Soil texture is loamy and sandy loamy. Carbonates were varied between 5.6% and 11.5%. The humus content in the solonchaks varied from 0.6% to 1%, and other soil contents from 0.2% to 1.3%. Soils in the dried seabed are saline and strongly saline. Salts are mostly concentrated in surface layer (0–20 cm) up to >16%. The chemistry of soil salinization is chloride-sulfate and sulfate-chloride. The soils are contaminated mainly with heavy metals as Cd and Zn.

Keywords Solonchaks · Sands · Saline soils · Humus · Soil texture · Carbonates · Heavy metals

4.1 Morphological Properties of the Soils

Flat plains and undulating ridged plains mainly represent the dried bed of the sea. There are 10 soil types in the region. Desert sands and sandy soils, sands with brown soil formation types, solonchaks (typical, takyr-like, coastal, and with saline marine sediments), desert brown and gray-brown soils, and meadow-floodplain-meadow soils were identified (Fig. 4.1). The soil profile of many samples is not strongly differentiated morphologically. Saline soils as solonchaks with different types are mainly common on the dried Aral seabed (Fig. 4.1).

In terms of morphological properties, the studied soils differ in vegetation cover and the thickness of the humus horizon. Effervescence from the surface that means soil is carbonate in surface.

Typical solonchaks occupy elevated relief elements and are found in combination with different soils. According to the mechanical composition (soil texture), they are variegated, layered, and of heavy granulometric composition throughout the

G. Issanova et al., *Soil Cover of the Dried Aral Seabed in Kazakhstan*,
https://doi.org/10.1007/978-3-031-29867-7_4

Fig. 4.1 Soil types in the dried Aral seabed

Fig. 4.2 Soil profile and its morphological characteristics of the typical solonchaks

entire depth of the soil profile. Solonchaks of sulfate and sulfate-chloride salinization are at the stage of modern salt accumulation.

Most of the soil surface is devoid of vegetation and covered with a continuous white coating of salt in the form of a warty brittle crust. The crust is easily pressed under the foot, which gets stuck in the underlying plump soil horizon.

Soil profile 11 of the *typical solonchaks* was taken in 2020, which has a rolling plain. Coordinates: N 45°81′247″; E 61°19′933″. H = 52 m a.s.l. Vegetation is saltwort-tamarisk and its height was from 10–15 cm to 1.5 m. Projective cover was 90%. A + B = 110 cm. Evaporation from the surface. No visible carbonates. Salt is from 8 to 50 cm depth, and 80–110 cm with veins of salts (Fig. 4.2).

Takyr-like solonchaks form a much less defined group of soils in comparison with other types of solonchaks. They are usually devoid of vegetation or very rare dying stunted biyurgun bushes are found. They differ from takyrs in a less dense surface, and under the soil crust, there is a scaly-powder-like that weakly compacted or loose horizon overflowing with salts. Efflorescence on the surface is insignificant, in the form of grayish-whitish thin spotted coatings that give the impression of being transparent.

The surface of takyr-like solonchaks is dense and even broken by cracks into drying polygons. The vegetation is very sparse and represented by halophytes.

Fig. 4.3 Soil profile and its morphological characteristics of the takyr-like solonchaks

Soil profile №7 of *takyr-like solonchaks* was taken in 2021, which has a plain and covered by salt-grass vegetation. Projective cover was 70–80%. Plant height was 20 cm and 1 m shrubs. The depth of the soil profile was 135 cm. Coordinates: N 45° 47′ 574″; E 61° 31′ 574″. H = 40 m a.s.l. (Fig. 4.3).

Coastal solonchaks are formed in the strip of relic seashore and confirmed by the initial stage of deflation. The solonchak process in these processes does not receive universal development. Wind activity and relief-forming processes significantly affect salt accumulation, creating conditions for horizontal and vertical redistribution of salts. The overdried layer formed from the surface of coastal solonchaks breaks the continuity of capillaries and weakens the processes of salt accumulation. Blowing and leaching of water-soluble salts begins along the positive elements of the relief, and salt accumulation continues along the blown basins. Therefore, coastal soils and coastal solonchaks with a wind-blown cover of small thickness, as a rule, form in combination with coastal solonchaks.

Soil profile № 4 of *coastal solonchaks* was taken in 2021, which has a rough plain and covered by sarsazan vegetation. Projective cover was 70%. Plant height was 30–50 cm and shrubs were 1 m. Soil profile depth was 120 cm. Coordinates: N 45° 18′ 757″; E 61° 21′ 419″. H = 46 m a.s.l. (Fig. 4.4).

Solonchaks with saline marine sediments from soil profile №19 (2021). It has a plain with lack of vegetation. Depth of the soil profile was 120 cm. Coordinates: N 44°40′ 543″; E 60°35′263″. H = 34 m a.s.l. (Fig. 4.5).

Fig. 4.4 Soil profile and its morphological characteristics of the coastal solonchaks

Deserts sands and sandy soils are widespread within the delta-alluvial plains of the Syrdarya river, especially in the northwestern part of the delta and on the coast of the Aral Sea. They are formed mainly as a result of the winding of ancient alluvium. The sands occupy vast areas in the region: in the north is the Aral Sea Karakum and in the south is Kyzylkum.

In our field expeditions, a number of sandy massifs were encountered: small hummocky formed by modern long processes of ancient alluvium winding; gently sloping-hilly and hilly-ridge island. The desert sands included soil profiles as following: 1, 2, 3, 11, 12 were taken in 2021 and soil profile 6 from 2020.

For instance, the soil profile 1 is located N 45° 25′ 132″ E 65° 15′ 380″ and H = 49 m a.s.l. It has a flat plain and zhuzgun-comb vegetation. Projective cover was 10%. The plant height was 1–2 m. The depth of the soil profile was 140 cm. A + B = 20 cm. The soil profile is poorly differentiated into horizons. A thin surface layer with 0.5–1.5 cm in grayish brittle crust stands out from the surface. Evaporation from the surface and salts are visible (Fig. 4.6).

Soil profile № 6 of *sands with brown soil formation types* was taken in 2021, and it has a flat plain. The vegetation is saltwort-comb and salt crystals are visible from the surface. Soil profile depth was 150 cm (Fig. 4.7).

Brown soils are the most common zonal soil types. They occupy automorphic landscapes of the inland deserts of the temperate zone. In the northern Aral Sea region, relatively young lacustrine marine and alluvial plains are equally distributed

Fig. 4.5 Soil profile and its morphological characteristics of the solonchaks with saline marine sediments

and composed by saline-layered sandy argillaceous deposits of marine transgressions and river deposits.

Soil profile 9 taken in 2020 is located on the watershed and has a rolling plain. The vegetation is ephemeral-wormwood with height of 5–10 cm. Projective cover was 60%. The soil depth was 90 cm. A + B = 28 cm. Evaporation from the surface. No visible carbonates. There is no visible salt. Coordinates: N 45°76′625″; E 61°31′009″. H = 63 m a.s.l. (Fig. 4.8).

The *gray-brown soils* of the central desert include the landscapes of the plains, predominantly of alluvial-accumulative origin. A common feature of the parent rocks of these soils is carbonate content and the presence of gypsum, and the content of carbonates often decreases with depth while gypsum increases.

The subzone of gray-brown soils is characterized by even greater aridity and continental climate compared to the brown soils of the northern desert. Gray-brown soils develop in automorphic landscapes and are characterized by a pronounced non-leaching type of water regime.

Soil profile 15 taken in 2021 is located in a plain. Vegetation cover is *Alhagi-Tamarix gallica* shrub and their height was from 30 cm to 1.5 m. Projective cover was 80%. The depth of the soil profile was 100 cm. Coordinates: N 45°28′44″; E 61°44′10″. H = 42 m a.s.l. (Fig. 4.9).

Fig. 4.6 Soil profile and its morphological characteristics of the desert sands

Meadow soil was taken from soil profile 16 in 2020. It has a flat plain. The vegetation is saltwort-tamarisk. Vegetation height was 10–15 cm; shrubs were 1–2 m. Projective cover was 40%. The soil profile depth was 100 cm. A + B = 20 cm. Carbonates and salts were not visible. Coordinates: N 46°73′93″; E 60° 88′89″. H = 38 m a.s.l. (Fig. 4.10).

Floodplain-meadow soils are formed under meadow-marsh vegetation and willows. The soil-forming rocks are layered and indistinctly layered alluvial deposits, mainly of loamy and clayey soil texture. They are found in the depressions of the near-channel and central floodplains, as well as on flat, low ridges and lowered, leveled spaces of the central floodplain. These soils are formed under conditions of periodic flooding by floodwaters, as well as under the influence of groundwater, which is within the soil profile for a long time. These soils are characterized by the processes of hydrogenous accumulation of iron and processes of gleying.

Soil profile 9 was taken in 2021 that has a plain. The vegetation cover is saltgrass-shrub and its height was 30 cm and 1.5 m for shrubs. Projective cover was 40–50%. The soil profile depth was 160 cm. Coordinates: N 45° 71′ 207″; E 61° 52′ 700″. H = 59 m a.s.l. (Fig. 4.11).

Fig. 4.7 Soil profile and its morphological characteristics of the sands with brown soil formation types

4.2 Chemical Composition of the Soils

Humus Content, Carbonates, and Acidity in Different Soil Types

The average humus content in the solonchaks upper 0–50 cm horizon varies from 0.6% to 1% in takyr-like and solonchaks with saline marine sediments, respectively (Fig. 4.12). However, the maximum humus was determined in the upper 0–10 cm soil horizon. The reaction of the soil environment is weakly alkaline (7.6–8.2). Carbonates are distributed through the soil profile and varied from 5.6% to 11.5%.

In *typical solonchaks*, the humus content in the upper 0–50 cm horizon varies from 0.55% to 0.94%. The humus maximum was determined in the upper 0–10 cm soil horizon. Carbonates are evenly distributed and varied in the soil profile from 5.61% to 6.43%. The reaction of the soil environment is alkaline (8.15–8.26). The absorption capacity is high (35.69–42.77 mg/eq), saturated more with sodium, less with calcium and magnesium (Table 4.1).

In *takyr-like solonchaks*, the humus content in the upper 0–50 cm horizon varies from 0.42% to 0.95%. The humus maximum was determined in the upper soil horizon. Carbonates are evenly distributed and varied in the soil profile from 6.56% to

Fig. 4.8 Soil profile and its morphological characteristics of the brown soils

7.08%. The reaction of the soil environment is alkaline (7.8–8.16). The absorption capacity is low (23.25–27.22 mg/eq) and saturated with calcium and then magnesium (Table 4.1).

In the *coastal solonchaks*, the humus content in the upper 0–50 cm soil horizon varies from 0.6% to 1.21%. The humus maximum was determined in the upper soil horizon. Carbonates are unevenly distributed and varied in the soil profile from 6.32% to 10.1%, with a maximum at the 80–100 cm horizon. The reaction of the soil environment is alkaline (7.7–8.19). The absorption capacity is high (41.74–45.29 mg/eq), significantly saturated with sodium, less calcium and magnesium (Table 4.1).

In *solonchaks with saline marine sediments*, the humus content in the upper 0–50 cm horizon varies from 0.53% to 1.32%. The humus maximum is set in the upper soil horizon. Carbonates varied in the soil profile from 7.49% to 11.5% and increased in 20–100 cm soil thick. The reaction of the soil environment is alkaline (7.5–7.84). The absorption capacity is high (12.06–39.73 mg/eq), significantly saturated with magnesium and sodium (Table 4.1).

In the *sands* (desert sands and sands with brown soil formation), average humus content varied 0.2–0.7%; pH was 7.1–7.5. CO_2 content was 6.7–8.2% (Fig. 4.13).

The content of humus in the *desert sands* upper soil horizon varies from 0.07% to 0.39%. The humus maximum is concentrated in the upper soil horizon. Carbonates

Fig. 4.9 Soil profile and its morphological characteristics of the gray-brown soils

are distributed along the soil profile almost evenly from 5.9% to 8.22%. The maximum is noted in the middle horizons. The reaction of the soil environment is slightly alkaline (7.24–7.42). Absorption capacity is low (3–6 mg/eq), saturated with calcium and then with magnesium (Table 4.2).

In *sands with brown soil formation types*, the humus content in the upper 0–50 cm horizon varied from 0.59% to 0.72%, slightly more than in desert sands. The humus maximum was determined in the upper soil horizon. Carbonates are unevenly distributed and varied in the soil profile from 4.43% to 11.46%. The maximum is noted from a half-meter horizon to a meter horizon. The reaction of the soil environment is slightly alkaline (7.31–7.78). The absorption capacity is low (3.46–5.45 mg/eq), saturated with magnesium and then with calcium (Table 4.2).

Average humus content varied between 0.6% and 0.8% in *floodplain-meadow and meadow soils*, respectively; the soil environment is weakly alkaline (7.7–8.0). Brown and gray-brown soils contain 0.5–0.6% humus, 1%–8.9% (CO_2), respectively, and soil was alkaline (7.6–7.8) (Fig. 4.14).

In *brown soils*, the humus content in the upper 0–50 cm horizon varies from 0.42% to 0.64%. The humus maximum is set in the upper soil horizon. Carbonates varied in the soil profile from 0.5% to 2.21%. The reaction of the soil environment is alkaline (7.5–8.1). The absorption capacity is low (3.46–8.41 mg/eq), saturated with calcium and magnesium (Table 4.3).

Fig. 4.10 Soil profile and its morphological characteristics of the meadow soils

In the *gray-brown soils*, the humus content in the upper 0–50 cm horizon varies from 0.61% to 0.67%. The humus maximum is set in the upper soil horizon. Carbonates are evenly distributed and varied in the soil profile from 7.59% to 9.44%. The reaction of the soil environment is alkaline (7.18–7.96). The absorption capacity is average (18.60–37.91 mg/eq), saturated with calcium, magnesium, and sodium (Table 4.3).

In *floodplain-meadow soils*, the humus content in the upper 0–50 cm horizon varies from 0.33% to 1.11%. The humus maximum is set in the upper soil horizon. Carbonates are evenly distributed and varied in the soil profile from 5.18% to 6.99%. The reaction of the soil environment is alkaline (7.5–7.9). The absorption capacity is high (16.78–53.24 mg/eq), saturated with sodium, potassium, and magnesium (Table 4.3).

The humus content, in *meadow soils*, in the upper 0–50 cm horizon varies from 0.6% to 1.25%. The humus maximum is set in the upper soil horizon. Carbonates are unevenly distributed, varied in the soil profile from 6.96% to 8.44%. The reaction of the soil environment is alkaline (7.8–8.2). The absorption capacity is average (12.96–24.62 mg/eq), and saturated with calcium and magnesium (Table 4.3).

Fig. 4.11 Soil profile and its morphological characteristics of the floodplain-meadow soils

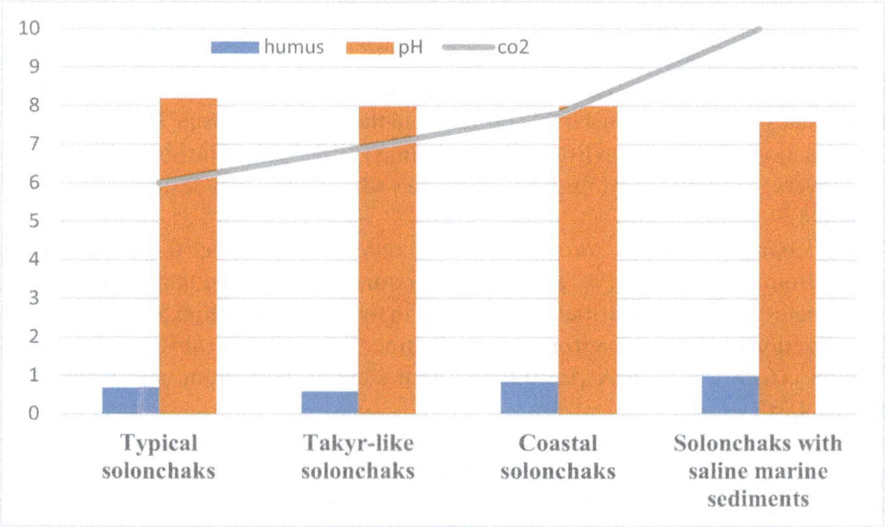

Fig. 4.12 Average content of humus, pH, and CO_2 in the solonchaks

Table 4.1 Chemical properties of the different kinds of solonchaks

Soil profile depth (cm)	pH	CO_2 (%)	Soil profile depth (cm)	Humus (%)	Absorbed bases (mg/eq/100 g of soil)			
					Na	K	Ca	Mg
Typical solonchaks								
0–20	8.15	5.61	0–10	0.94	19.74	0.87	12	10.16
20–50	8.14	6.43	10–20	0.71	13.69	0.49	10.16	12
50–80	8.26	5.88	20–50	0.55	14.71	0.24	8.82	11.76
80–100	8.3	6						
Takyr-like solonchaks								
0–20	7.8	6.92	0–10	0.95	3.77	0.42	12.14	6.92
20–50	8	7.08	10–20	0.53	5.34	0.47	11	8.48
50–80	8.16	7.06	20–50	0.42	9.49	0.4	7.11	10.22
80–100	8.02	6.56						
100–140	7.8	6.74						
Coastal solonchaks								
0–20	7.75	6.35	0–10	1.21	18.37	0.78	14.7	10.72
20–50	8.01	8.73	10–20	0.74	17.28	0.55	20.17	7.29
50–80	7.95	7.56	20–50	0.6	21.68	0.42	8.65	10.99
80–100	8.19	10.1						
100–140	7.91	6.32						
Solonchaks with saline marine sediments								
0–20	7.73	7.49	0–10	1.32	49.32	1.97	10.23	97.43
20–50	7.63	11.5	10–20	1.18	11.29	0.16	18.48	18.31
50–80	7.84	10.9	20–50	0.53	38.18	0.07	14.36	20.62
80–100	7.51	10.16						
100–140	7.51	10.6						

Salts in the Soils and Salt Profiles

The *solonchaks with saline marine sediments* are highly saline (>16%). Salts in the typical solonchaks (2.4%) and solonchaks with saline marine sediments are mostly concentrated in the surface soil layer (0–20 cm), in the takyr-like and coastal solonchaks through the soil profile (Fig. 4.15). Among the cations, the sodium cation (Na^+) prevailed and ranged from 5.13 to 188.6 mg/eq/100 g of soil. Sulfate ions (SO_4^-) and chloride ions (Cl^-) prevailed among the anions, and they ranged from 8.42 to 125.3 meq/100 g and 8.86 to 24.98 meq/100 g of soil, respectively. The chemistry of salinization is chloride-sulfate and sulfate-chloride.

In the aqueous extract of the *typical solonchaks*, the amount of anions was greater than that of cations. Among the cations, the sodium cation (Na^+) prevailed; on a 140 cm soil profile, it ranged from 5.13 to 20.42 mg/eq/100 g of soil. High content was noted on the upper 0–20 cm soil horizon. Chloride ions (Cl^-) prevailed among the anions and they ranged from 8.86 to 24.98 meq/100 g of soil. High contents were noted on the upper horizons. The chemistry of salinization is sulfate-chloride.

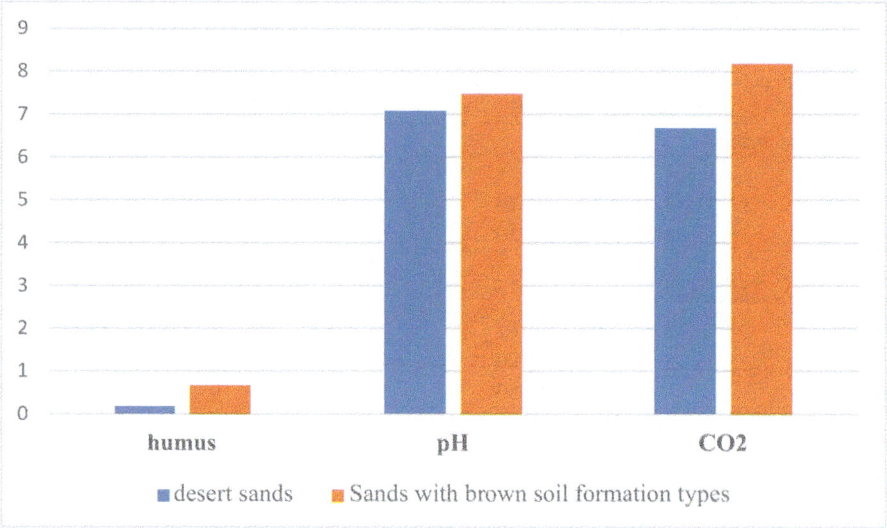

Fig. 4.13 Average content of humus, pH, and CO_2

Table 4.2 Chemical properties of the sands

Soil profile depth (cm)	pH	CO_2 (%)	Soil profile depth (cm)	Humus (%)	Absorbed bases (mg/eq/100 g of soil)			
					Na	K	Ca	Mg
Desert sands								
0–20	7.42	6.61	0–10	0.39	0.14	0.12	1.9	1.41
20–50	7.29	6.73	10–20	0.08	0.1	0.19	4.84	1.65
50–80	7.24	8.22	20–50	0.07	0.09	0.08	2.88	1.24
80–100	7.27	5.9						
100–140	7.38	6.25						
Sands with brown soil formation types								
0–20	7.54	7.99	0–10	0.72	3.21	0.16	6.43	12
20–50	7.5	7.84	10–20	0.67	4.98	0.31	7.18	9.77
50–80	7.78	11.46	20–50	0.59	4.05	0.14	2.06	7.59
80–100	7.77	9.3						
100–140	7.31	4.43						

In *typical* solonchaks, in terms of the amount of salts (%), the upper 0–20 cm horizon is solonchak, 20–50 cm and 80–100 cm horizons are highly saline, and 50–80 cm are medium saline (Table 4.4).

According to the salt profile, it belongs to the solonchak profile. The maximum water-soluble salts are accumulated in the upper part of the profile with a significant amount of salts throughout the profile. This indicates the stage of progressive salinization (Fig. 4.16).

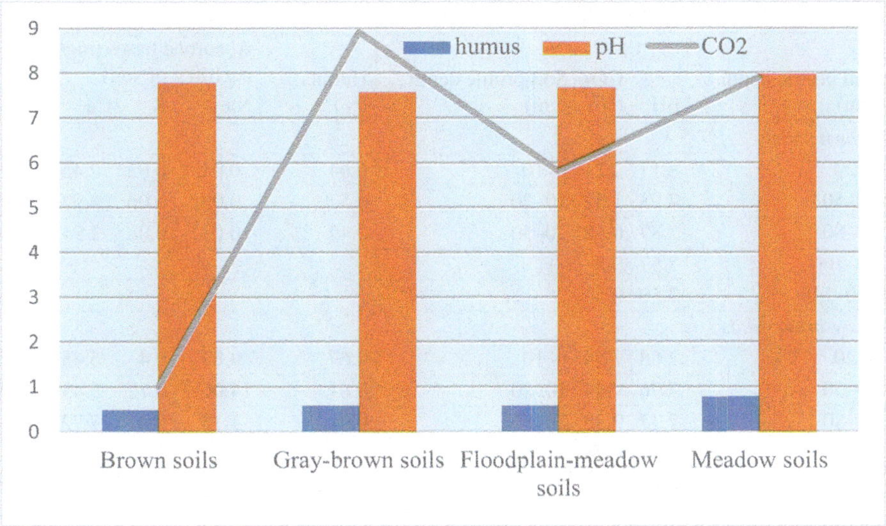

Fig. 4.14 Average content of humus, pH, and CO_2

In the water extract of *takyr-like solonchaks*, the amount of anions was greater than cations. Among the cations, the sodium cation (Na^+) prevailed on a 140 cm soil profile, and it ranged from 11.95 to 17.88 mg/eq/100 g of soil. Sodium cations are distributed evenly along the soil profile. Sulfate ions (SO_4^-) prevailed among the anions and they ranged from 8.42 to 13.92 meq/100 g of soil, high contents throughout the soil profile. The chemistry of salinization is chloride-sulfate (Fig. 4.16). In takyr-like solonchaks, the entire soil profile (0–140 cm) was highly saline in terms of the amount of salts (%). Salt profile refers to a prism-shaped salt profile in which the salts are distributed relatively evenly throughout the profile.

The sum of anions and cations in the *coastal solonchaks* was at an equal level. Among the cations, the sodium cation (Na^+) prevailed on a 140 cm soil profile and it ranged from 11.25 to 23.51 mg/eq/100 g of soil. High contents of sodium cations are noted in the upper 0–20 cm and 50–80 cm horizons. Sulfate ions (SO_4^-) prevailed among the anions and they ranged from 11.49 to 20.79 meq/100 g of soil. High contents were observed in the first half-meter layer of the soil profile. The chemistry of salinization is chloride-sulfate (Fig. 4.16). In the coastal solonchaks, in terms of the amount of salts (%), the upper 0–20 cm and 50–80 cm horizons are solonchak, and 80–100 cm and 100–140 cm horizons are highly saline. The salt profile refers to a salt profile with multiple peaks. It indicates that long periods of salinization during the formation of the soil were replaced by periods of desalinization.

The total amount of cations in the *solonchaks with saline marine sediments* was greater than anions. Among the cations, the sodium cation (Na^+) prevailed on a 140 cm soil profile (Fig. 4.16). It ranged from 48.45 to 188.60 mg/eq/100 g of soil. Very high contents of sodium cations are noted in the upper 0–20 cm horizon. Sulfate ions (SO_4^-) prevailed among the anions. They ranged from 37.9 to

Table 4.3 Chemical properties of the soils

Soil profile depth (cm)	pH	CO_2 (%)	Soil profile depth (cm)	Humus (%)	Absorbed bases (mg/eq/100 g of soil)			
					Na	K	Ca	Mg
Brown soils								
0–20	8.17	2.21	0–10	0.64	0.01	0.02	2.45	0.98
20–50	7.93	1.52	10–20	0.57	0.03	0.06	3.92	1.96
50–80	7.87	0.68	20–50	0.42	0.01	0.08	2.94	5.39
80–100	7.57	0.5						
100–140	7.50	0.5						
Gray-brown soils								
0–20	7.68	7.59	0–10	0.67	0.87	0.4	5.45	11.88
20–50	7.96	9.44	10–20	0.61	14.06	0.12	5.45	12.38
50–80	7.37	9.28	20–50	0.61	4.18	0.07	27.72	5.94
80–100	7.77	9.28						
100–140	7.18							
Floodplain-meadow soils								
0–20	7.8	5.62	0–10	1.11	19.73	1.05	11.09	21.37
20–50	7.95	6.99	10–20	0.5	3.08	0.11	7.02	11.84
50–80	7.82	5.35	20–50	0.33	2.32	0.06	5.96	8.44
80–100	7.95	5.18						
100–140	7.58	5.64						
Meadow soils								
0–20	7.97	6.96	0–10	1.25	1.11	0.30	4.65	7.84
20–50	7.89	8.2	10–20	0.69	1.18	0.27	4.9	6.61
50–80	8.22	7.93	20–50	0.6	0.91	0.69	14.45	8.57
80–100	8.27	8.44						

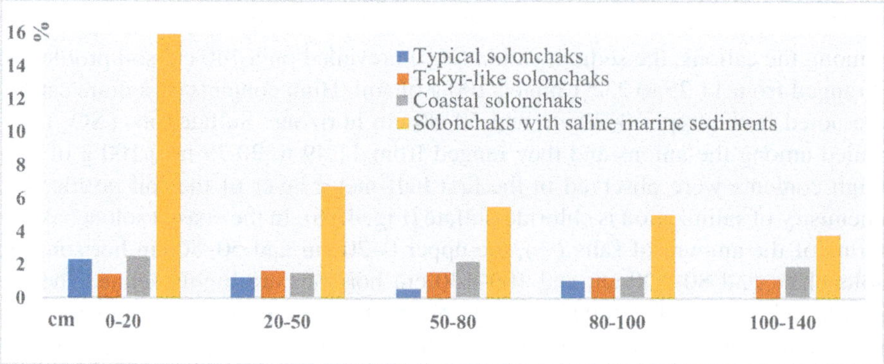

Fig. 4.15 Salt content in different kinds of solonchaks

Table 4.4 Water extract of the different kinds of solonchaks

Soil profile depth (cm)	CO_3^-	HCO_3^-	Cl^-	SO_4^-	Ca^+	Mg^{2+}	Na^+	K^+	Soil salinity (%)
	mmol (eq)/100 g								
Typical solonchaks									
0–20	0.04	0.33	24.98	15.09	11.57	7.77	20.42	0.72	2.448
20–50	0.04	0.2	18.61	4.22	5.4	5.6	11.95	0.32	1.344
50–80	0.04	0.35	18.61	0.22	1.78	3.37	5.13	0.09	0.572
80–100	0.04	0.33	8.36	3.78	4.41	3.52	11.34	0.21	1.14
Takyr-like solonchaks									
0–20	0.00	0.390	8.49	13.92	8.4	2.74	12.37	0.5	1.53
20–50	0.04	0.37	10.04	10.53	3.4	3.31	17.88	0.3	1.68
50–80	0.04	0.32	8.93	10.83	2.47	4.22	14.88	0.3	1.55
80–100	0.04	0.36	10.41	8.42	2.88	3.78	11.95	0.28	1.34
100–140	0.00	0.45	4.4	5.00	0.4	0.6	10	0.11	1.20
Coastal solonchaks									
0–20	0.003	0.3	17.06	20.79	10.96	4.44	23.51	0.66	2.45
20–50	0.04	0.44	7.27	14.5	7.0	2.25	13.5	0.56	1.48
50–80	0.00	0.45	14.49	19.6	9.73	3.73	20.92	0.85	2.24
80–100	0.00	0.37	7.33	11.49	5.63	3.23	11.25	0.38	1.26
100–140	0.00	0.32	11.66	17.65	7.53	4.8	17.9	0.54	1.92
Solonchaks with saline marine sediments									
0–20	0.267	0.372	127.33	125.3	18.244	57.822	188.606	3.963	16.14
20–50	0.3	0.383	26.8	74.1	6.733	17.067	77.177	2.992	6.77
50–80	0.2	0.3	23.467	60.5	7.2	12.867	63.009	1.77	5.58
80–100	0	0.383	21.867	53.7	4.333	11.8	61.889	1.852	5.11
100–140	0	0.383	20.4	37.9	3.667	10.333	48.458	1.351	3.93

125.3 meq/100 g of soil; high contents were observed in the first half-meter layer of the soil profile. The chemistry of salinization is chloride-sulfate. In the solonchaks with saline marine sediments, in terms of the amount of salts (%), the entire soil profile is solonchak, and it refers to the salt profile with a pronounced maximum (high contents) of salts observed in the 0–20 cm soil horizon, while the salt content in the rest of the soil profile is low, indicating the initial stage of salinization.

A large amount of sodium sulfate in the composition of salts is the result of secondary colloidal exchange reactions. Analysis shows that the average salt content in the 0–50 cm layer of marine sediments is 120 t/ha, and on the entire exposed surface was 163 million tons. If we imagine that these salts will be carried out during aeolian processing of sediments by prevailing west-east winds to a distance of up to 200–400 km along the front long axis of the sea, then this will amount to 20 t/ha of aeolian salt inflow (Faizov and Tapalova 2003). This is a real danger to the environment of the Aral Sea region and its population. The current sanitary and epidemiological situation in the region is characterized by an extremely high level of infectious and noninfectious morbidity, an increased level of oncological, allergic

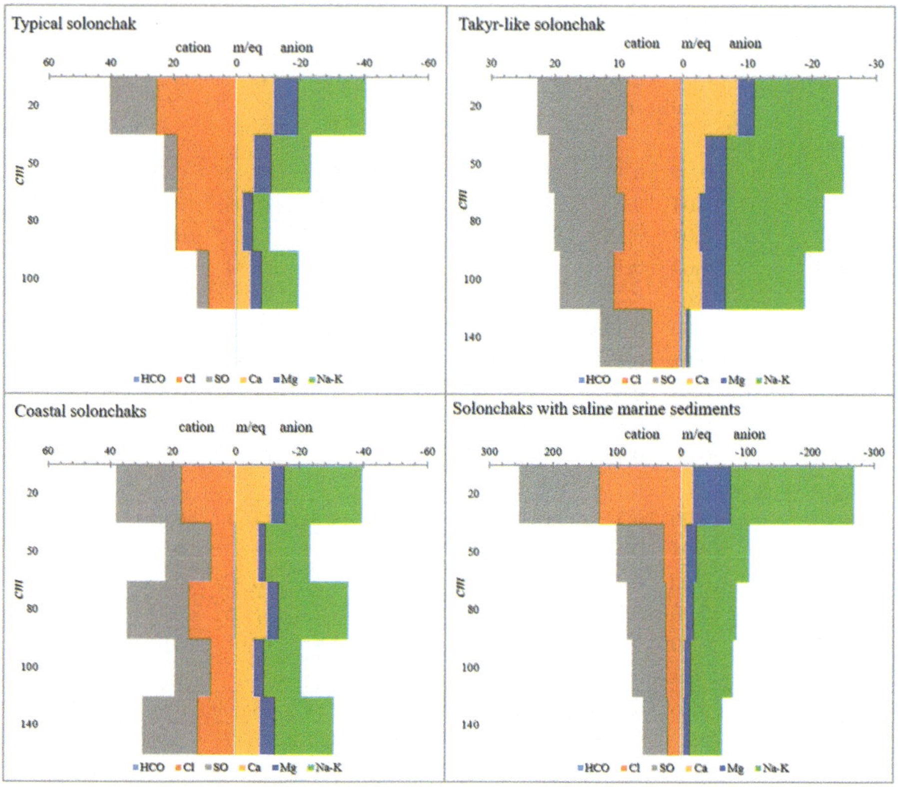

Fig. 4.16 Salt profiles of different kinds of solonchaks

and genetic diseases, which is extremely unfavorable for the demographic situation in the region.

Surface layers of the *desert sands* are nonsaline; however, lower layers (80–140 cm) are slightly saline. Salt content is high in the sands with brown types of soil formation. Surface layer (0–20 cm) is very strongly saline (>2%) (Fig. 4.17).

In the aqueous extract of *desert sands*, the amount of cations was greater than that of anions. Among the cations, the calcium cation (Ca^{+}) prevailed; on a 140 cm soil profile, it ranged from 0.465 to 4.52 mg/eq/100 g of soil. Marked high content from 80 cm to the depth of the soil profile. Sulfate ions (SO_4^-) prevailed among the anions. They ranged from 0.29 to 4.86 meq/100 g of soil, also with high contents in the lower 80–140 cm horizons (Table 4.5). The chemistry of salinization is chloride-sulfate. By the amount of salts (%) in the desert sands, the upper 3 horizons and 0–80 cm are nonsaline, and from 80 to 140 cm are slightly saline.

The *salt profile* of the *desert sands* refers to the pyramidal salt profile, in which the salt content increases with depth with a significant amount of salt throughout the profile and the presence of one salt maximum, confined to the lower horizons

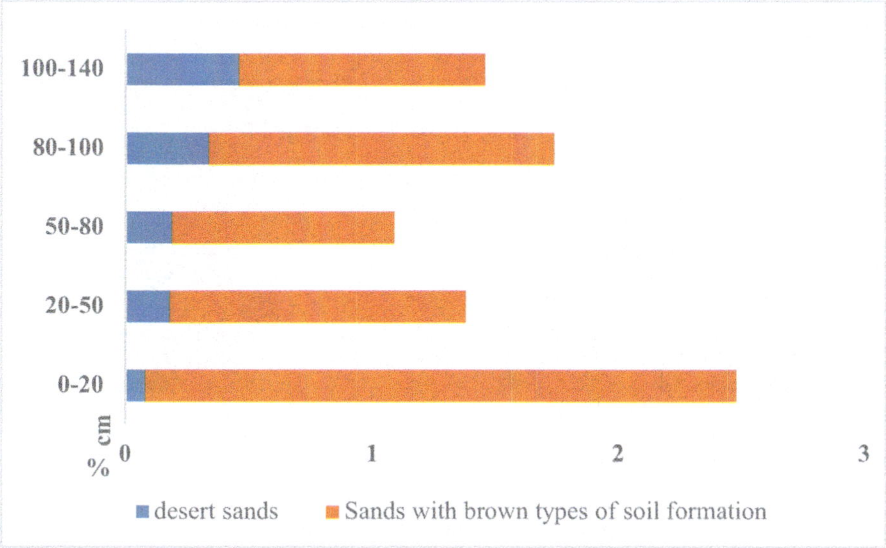

Fig. 4.17 Salt content of the desert sands and its brown types of soil formation

Table 4.5 Water extract of desert sands

Soil profile depth (cm)	CO_3^-	HCO_3^-	Cl^-	SO_4^-	Ca^+	Mg^{2+}	Na^+	K^+	Soil salinity (%)
	mmol (eq)/100 g								
Desert sands									
0–20	–	0.42	0.274	0.29	0.465	0.235	0.13	0.113	0.08
20–50	–	0.378	0.348	0.53	0.552	0.302	0.097	0.171	0.18
50–80	–	0.353	0.974	1.98	0.502	0.803	0.318	0.352	0.19
80–100	–	0.437	1.18	4.0	2.134	1.36	0.990	0.409	0.34
100–140	–	0.453	1.472	4.86	4.52	0.7	0.865	0.414	0.46
Sands with brown types of soil formation									
0–20	–	0.425	18.640	19.8	10.670	7.075	15.921	1.669	2.37
20–50	–	0.463	6.240	11.3	4.900	2.25	10.648	0.449	1.18
50–80	0.100	0.65	7.050	6.8	2.300	1.95	10.205	0.296	0.94
80–100	0.200	0.383	11.707	9.8	2.800	3.867	15.121	0.374	1.37
100–140	0.200	0.383	9.067	7.7	1.800	2.8	12.681	0.338	1.09

(Fig. 4.18). This indicates the development of drainage, subsidence of groundwater, and removal of various salts, their removal into groundwater.

A salt profile of the *sands with brown types of soil formation* refers to a salt profile with multiple peaks. It indicates that long periods of salinization during the formation of the soil were replaced by periods of desalinization (Fig. 4.18).

In the water extract of *sands with brown types of soil formation*, the sum of anions predominated over cations. Among the cations, the sodium cation (Na^+)

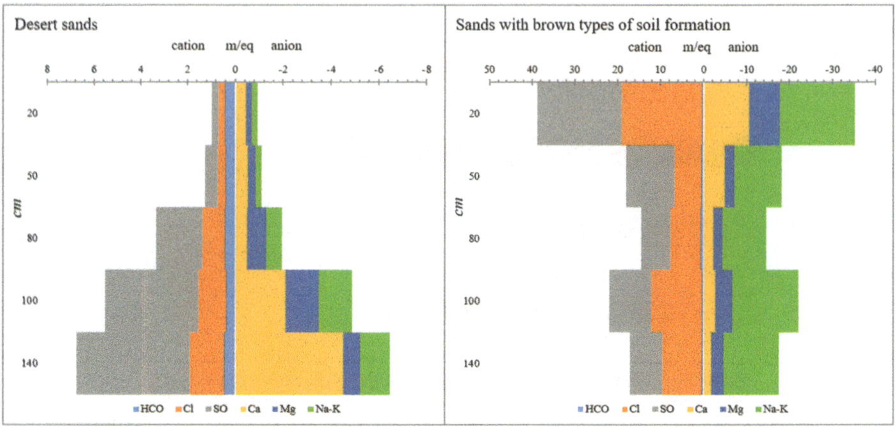

Fig. 4.18 Salt profiles of the desert sands and its brown types of soil formation

prevailed; on a 140 cm depth of the soil profile, it ranged from 10.648 to 15.921 mg/eq/100 g of soil (Table 4.5). Sodium cations are distributed evenly along the soil profile. Sulfate ions (SO_4^-) prevailed among the anions, and they ranged from 6.8 to 19.8 meq/100 g of soil. High contents were observed in the first half-meter layer of the soil profile. The chemistry of salinization is chloride-sulfate. In terms of the amount of salts (%), the upper 0–20 cm horizon is solonchak, and the remaining horizons down to the depth (140 cm) of the soil profile are highly saline.

Gray-brown soils are highly saline and in surface layer (0–20) salts >19% compared with brown desert soils. Furthermore, floodplain-meadow soils are also saline (3.4%) and meadow soils are more saline (1.2%) in deep soil horizon at 100–140 cm (Fig. 4.19). The amounts of cations and anions in the *gray-brown soils* were almost equal. Among the cations, the sodium cation (Na^+) prevailed on a 140 cm soil profile. It ranged from 68.692 to 179.09 mg/eq/100 g of soil. The highest concentrations of sodium cations were noted in the upper 0–20 cm horizon; sodium cations were also high along the soil profile. Chloride ions (Cl^-) prevailed among the anions and they ranged from 21.200 to 174.00 meq/100 g of soil. High contents were determined on the upper 0–20 cm horizon of the soil profile. The chemistry of salinization is sulfate-chloride (Table 4.6).

By the amount of salts (%) in gray-brown soils, the entire soil profile is solonchak, with salt maxima in the upper (0–20 cm) and middle (50–80 cm) horizons. The salt profile refers to the salt profile with a pronounced maximum of salts in the upper horizon, while the salt content in the rest of the soil profile is low, indicating the initial stage of salinization (Fig. 4.20).

In the *brown soils*, the sum of anions and cations was at the same level. Among the cations, the calcium cation (Ca^+) prevailed, fluctuating from 0.99 to 0.3 mg/eq/100 g of soil per 100 cm soil profile. Sulfate ions (SO_4^-) prevailed among the anions and they ranged from 0.045 to 1.8 meq/100 g of soil; high contents were observed below 20 cm of the soil profile. The chemistry of salinity is sulfate.

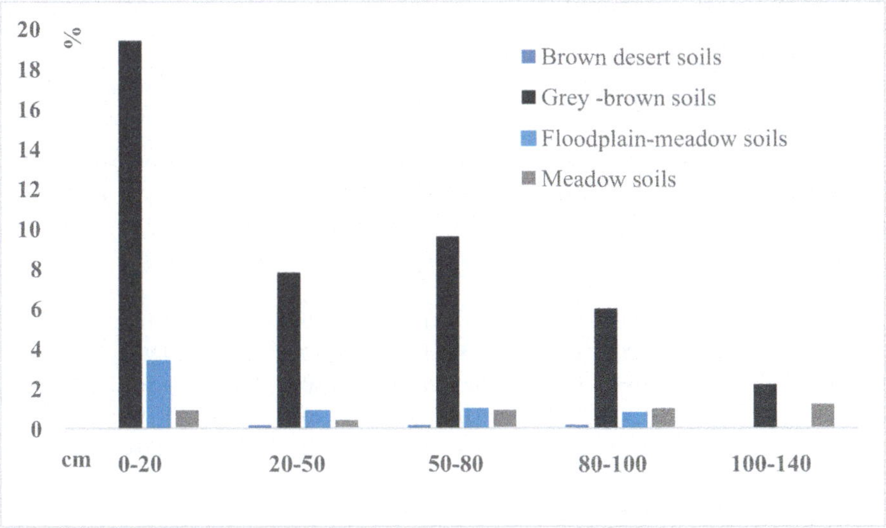

Fig. 4.19 Salt content of the desert sands

According to the amount of salts (%) in brown soils, all considered horizons are nonsaline (Table 4.6).

The salt profile with the maximum salt content at a certain depth against the background of desalinization of the upper horizons indicates the formation of an illuvial horizon in relation to the soil layers (Fig. 4.20).

In the *meadow soils*, the sum of anions prevailed over the sum of cations. Among the cations, the sodium cation (Na^+) prevailed on a 140 cm soil profile. It ranged from 0.23 to 11.54 mg/eq/100 g of soil. High contents of sodium cations were noted in the upper 0–20 cm horizon and in the lower 80–140 cm horizons. Sulfate ions (SO_4^-) prevailed among the anions and they ranged from 11.49 to 20.79 meq/100 g of soil; high contents were observed in the first half-meter layer of the soil profile. The salinity chemistry is mixed (Table 4.6).

According to the total salts (%) in meadow soils, the upper horizons up to 80 cm are weakly saline, and from 90 cm to 140 cm, they are strongly saline (Fig. 4.21). A salt profile refers to a salt profile with multiple peaks. It indicates that long periods of salinization during soil formation were replaced by periods of desalinization.

In the water extract of *floodplain-meadow soils*, the amount of cations was slightly greater than the amount of anions. Among the cations, the sodium cation (Na^+) prevailed on a 100 cm soil profile and it ranged from 1.98 to 37.48 mg/eq/100 g of soil. High contents of sodium cations were noted in the upper 0–20 cm soil horizon. Sulfate ions (SO_4^-) prevailed among the anions and they ranged from 7.47 to 35.35 meq/100 g of soil; high contents were determined on the upper 0–20 cm horizon of the soil profile. The chemistry of salinization is chloride-sulfate (Table 4.6). By the amount of salts (%) in the floodplain-meadow soils, the 0–20 cm horizon is solonchak, the 50–80 cm horizon is highly saline, and the 20–50 cm and 80–100 cm are moderately saline. The salt profile refers to the salt profile with a

Table 4.6 Water extract of the soils

Soil profile depth (cm)	CO_3^-	HCO_3^-	Cl^-	SO_4^-	Ca^+	Mg^{2+}	Na^+	K^+	Soil salinity (%)
	mmol (eq) /100 g								
Brown desert soils									
0–20	0.04	0.58	0.06	0.045	0.99	0.2	0.06	0.05	0.054
20–50	0.04	0.52	0.08	1.8	0.3	0.5	0.09	0.17	0.169
50–80	0.04	0.52	0.08	1.8	0.99	0.69	0.49	0.23	0.169
80–100	0.04	0.48	0.42	1.37	0.99	0.59	0.6	0.1	0.155
Gray-brown soils									
0–20	0.350	0.400	174.000	145.8	18.000	105.000	179.091	4.160	19.40
50–50	0.200	0.200	34.000	82.5	11.200	16.400	85.677	4.971	7.78
50–80	–	0.450	31.200	109.9	4.400	28.800	111.153	4.863	9.60
80–100	–	0.300	21.200	69.0	6.000	16.200	68.692	1.196	6.03
100–140	–	0.350	8.800	24.0	1.900	5.200	26.230	0.510	2.23
Floodplain-meadow soils									
0–20	0.1	0.408	20.57	35.35	7.94	12.45	37.48	0.64	3.39
20–50	0.03	0.435	6.96	7.95	7.76	4.59	1.98	0.33	0.92
50–80	0.046	0.477	2.55	10.07	4.07	3.53	8.44	0.14	1.034
80–100	0	0.412	2.3	7.47	2.8	2.88	6.38	0.19	0.79
Meadow soils									
0–20	0	0.46	5.31	7.15	0.99	2.22	7.35	0.3	0.86
20–50	0.04	0.52	0.31	4.3	0.5	0.74	0.36	0.025	0.4
50–80	0.04	0.56	0.23	7.18	0.99	1.73	0.23	0.2	0.87
80–100	0.04	0.28	15.57	2.64	1.49	3.71	11.54	0.36	1.07
100–140	0.04	0.20	10.50	2.60	1.20	3.10	10.20	0.26	1.20

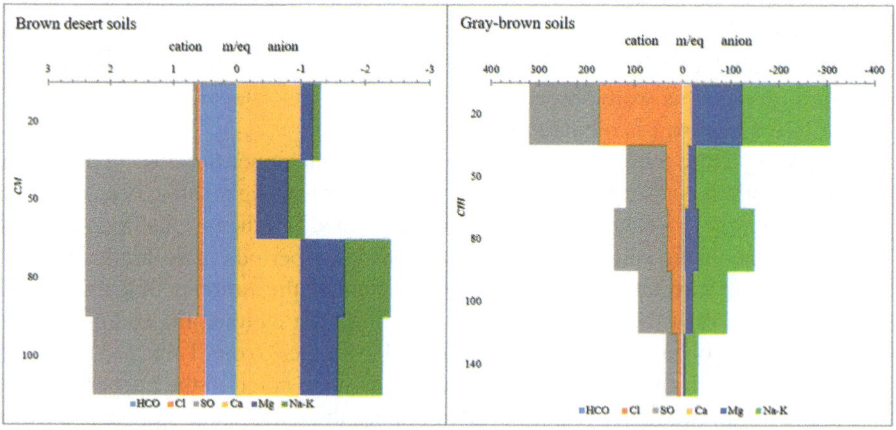

Fig. 4.20 Salt profiles of the brown and gray-brown soils

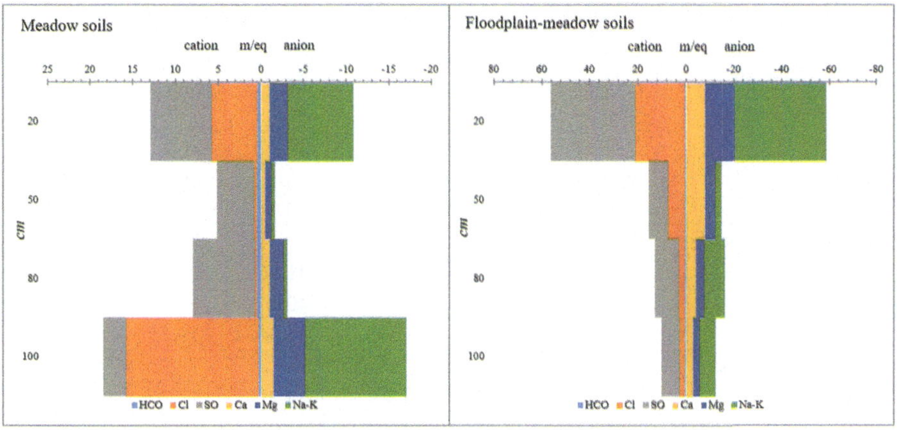

Fig. 4.21 Salt profiles of the meadow and floodplain-meadow soils

pronounced maximum of salts in the upper horizon, while the salt content in the rest of the soil profile is low, indicating the initial stage of salinization (Fig. 4.21).

Heavy Metals in the Soils

The content of heavy metals in 0–50 cm depth is various. The soils in the dried Aral seabed are contaminated by heavy metals such as cadmium (Cd) and zinc (Zn). In solonchaks (*typical, takyr-like, coastal*), cadmium (Cd) exceeds the threshold limit value (TLV) from 15 to 21 times and ranged from 5.5 to 8.8 mg/kg. Zinc (Zn) in the *solonchaks with saline marine sediments* exceeds the TLV by 0.8 times.

In the soil profile of *typical solonchaks*, the total content of heavy metals is distributed in the following (mg/kg): Zn, 30.6–36.1; Cu, 18–19.8; Cd, 6.1–8.8 (Table 4.7). Among the studied heavy metals, the gross content of cadmium exceeds the TLV by 17.6 times and corresponds to the clarks in the lithosphere. The content of other heavy metals does not exceed MPC and below their clarks in the lithosphere.

The following results were obtained for mobile heavy metals forms (mg/kg): Zn, 3.32–4.95; Cu, 0.72–1.22; Cd, not found (Table 4.7).

The total content of heavy metals in the *takyr-like solonchaks* is distributed in the following order (mg/kg): Zn, 40.9–44.06; Cu, 9.9–11.75; Cd, 5.2–5.56. Among the studied heavy metals, the content of cadmium exceeds the TLV by 21.12 times; the contents of other HMs do not exceed TLV and are below their clarks in the lithosphere.

The mobile forms of HM obtained the following results (mg/kg): Zn, 3.13–3.58; Cu, 1.55–2.14; Cd, 0.96–1.33. Mobile forms of Cd are 2.66 times higher than the TLV of mobile forms of cadmium.

Table 4.7 Mobile and gross forms of heavy metals in different kinds of solonchaks

	Mobile			Gross		
Soil profile depth (cm)	Zn	Cu	Cd	Zn	Cu	Cd
Typical solonchaks						
0–10	3.32	0.72	–	31.1	16.3	6.8
10–20	4.95	1.12	–	36.1	19.8	6.1
20–50	4.37	1.22	–	30.6	18	8.8
Takyr-like solonchaks						
0–10	3.58	1.55	1.26	41.66	9.9	5.56
10–20	3.13	1.75	1.33	40.9	10.12	5.2
20–50	3.46	2.14	0.96	44.06	11.75	5.6
Coastal solonchaks						
0–10	5.82	2.82	1.45	35.9	11.7	5
10–20	6.47	3.65	1.5	21.9	12.2	7.9
20–50	5.17	5.72	0.95	20.5	11.6	7.5
Solonchaks with saline marine sediments						
0–10	3.58	3.33	1.21	23.53	11.13	0.86
10–20	4.23	3.63	1.43	40.26	14.33	2.4
20–50	5.03	3.9	1.43	64.26	20.8	2.26

The *coastal solonchaks* content of heavy metals are in the following way (mg/kg): Zn, 20.5–35.9; Cu, 11.6–12.2; Cd, 5–7.9. Among the studied heavy metals, the content of cadmium exceeds the TLV by 15.8 times. The contents of other HMs do not exceed TLV and are below their clarks in the lithosphere.

The following results were obtained for mobile HM forms (mg/kg): Zn, 5.17–6.47; Cu, 2.82–5.72; Cd, 0.95–1.5. Mobile forms of Cd are 3 times higher than the TLV of mobile forms of cadmium.

In the soil profile of *solonchaks with saline marine sediments*, the total content of heavy metals is distributed in the following way (mg/kg): Zn, 23.53–64.26; Cu, 11.13–20.8; Cd, 0.86–2.26. Among the studied heavy metals, cadmium exceeds the TLV by 4.52 times and zinc by 0.8 times. The content of these elements is closer to their clarks in the lithosphere. The content of other HMs does not exceed TLV and below their clarks in the lithosphere.

The mobile forms of HM obtained the following results (mg/kg): Zn, 3.58–5.03; Cu, 3.33–3.9; Cd, 1.21–1.43. Mobile forms of Cd are 2.86 times higher than the TLV of mobile forms of cadmium.

Desert sands were also contaminated by cadmium (Cd) and it exceeds the TLV by 2–3 times and ranged from 0.66 to 1.6 mg/kg.

In the soil profile of *desert sands*, the gross content of heavy metals is distributed in the following way (mg/kg): Zn, 13.2–15.26; Cu, 4.93–5.73; Cd, 0.66–1.13 (Table 4.8). Among the studied heavy metals, the content of cadmium exceeds the TLV by two times. In addition, the content of other heavy metals does not exceed TLV. All elements are below their clarks in the lithosphere. The following results were obtained for mobile heavy metals forms (mg/kg): Zn, 1.2–1.45; Cu, 0.8–1.21;

Table 4.8 Mobile and gross forms of heavy metals in desert sands (mg/kg)

	Mobile			Gross		
Soil profile depth (cm)	Zn	Cu	Cd	Zn	Cu	Cd
Desert sands						
0–10	1.42	0.8	0.35	15.26	4.93	0.66
10–20	1.2	0.81	0.37	13.2	5.2	1.13
20–50	1.45	1.21	0.31	15.26	5.73	1.2
Sands with brown types of soil formation						
0–10	2.8	1.52	1	25.2	8.9	1.6
10–20	2.57	2.8	0.8	28.3	13.2	0.6
20–50	2.47	2.3	0.8	26	13.3	0.3

Table 4.9 Mobile and gross forms of heavy metals in the brown and gray-brown soils (mg/kg)

	Mobile			Gross		
Soil profile depth (cm)	Zn	Cu	Cd	Zn	Cu	Cd
Brown soils						
0–10	0.2	0	0	28.8	4.8	11.2
10–20	0.1	0	0	34	6.8	18.4
20–50	3.3	0	0	19.2	3.6	8
Gray-brown soils						
0–10	5.8	0.8	1.2	49.6	27.6	2
10–20	6	0.8	1.5	87.2	28.8	0.8
20–50	22.3	2.5	0.7	52.4	30.4	0

Cd, 0.31–0.37. The mobile forms of all heavy metals are below the TLV of these elements.

In the soil profile of the *sands with brown types of soil formation*, the total content of heavy metals is distributed in the following (mg/kg): Zn, 25.2–28.3; Cu, 8.9–13.3; Cd, 0.3–1.6. Among the studied heavy metals, the content of cadmium exceeds the TLV by 3 times. The content of other heavy metals does not exceed TLV. All elements are below their clarks in the lithosphere.

The mobile forms of heavy metals obtained the following results (mg/kg): Zn, 2.47–2.8; Cu, 1.52–2.8; Cd, 0.8–1.0. Mobile forms of Cd are 2 times higher than TLV.

In the *brown soils*, Cd was exceeding the TLV by 36 times and 4 times in the *gray-brown soils*.

In the soil profile of *brown soils*, the total content of heavy metals is distributed in the following way (mg/kg): Zn, 19.2–34; Cu, 3.6–6.8; Cd, 8–18.4 (Table 4.9). Among the studied heavy metals, cadmium (Cd) exceeds the TLV by 36 times; the content of cadmium is higher than its clarks in the lithosphere. The content of other HMs does not exceed TLV and below their clarks in the lithosphere.

The following results were obtained for mobile HM forms (mg/kg): Zn, 0.1–3.3; Cu, 0; Cd, 0.

In the soil profile of *gray-brown soils*, the total content of heavy metals is distributed in the following way (mg/kg): Zn, 49.6–87.2; Cu, 27.6–30.4; Cd, 0.8–2.0. Among the studied heavy metals, cadmium exceeds the TLV by 4 times. The contents of all HMs do not exceed below their clarks in the lithosphere.

The following results were obtained for mobile HM forms (mg/kg): Zn, 5.8–22.3; Cu, 0.8–2.5; Cd, 0.7–1.5 (Table 4.9). Mobile forms of Cd are 3 times higher than the TLV of mobile forms of cadmium.

Cadmium (Cd) and zinc (Zn) was also exceeding the TLV by 29 times and 0.8 times in the *meadow soils*, respectively, and by 28 times and 1.4 times and in the *floodplain-meadow soils.*

In the soil profile of *meadow soils*, the total content of heavy metals is distributed in the following way (mg/kg): Zn, 27.8–67.4; Cu, 8–15.2; Cd, 8.4–14.8. Among the studied heavy metals, cadmium exceeds the TLV by 29.6 times and zinc by 0.8 times. Among HMs, cadmium and zinc are higher than their clarks in the lithosphere (Table 4.10).

The mobile forms of HM obtained the following results (mg/kg): Zn, 3.5–6.55; Cu, 2.45–3.35; Cd, 0. Mobile forms of Cu are 0.25 times higher than the TLV of mobile forms.

In the soil profile of *floodplain-meadow soils*, the total content of heavy metals is distributed in the following way (mg/kg): Zn, 36.7–62.7; Cu, 14.5–19.7; Cd, 11.2–14.1. Among the studied heavy metals, cadmium exceeds the TLV by 28.2 times and zinc by 1.4 times. Among HMs, cadmium and zinc are higher than their clarks in the lithosphere.

The following results were obtained for mobile HM forms (mg/kg): Zn, 4.12–10.47; Cu, 1.95–3.1; Cd, 0.75–1.5. Mobile forms Cd are 3 times higher than TLV of mobile forms (Table 4.10).

The main source of soil contamination with cadmium can be explained by industrial emissions and wastewater. In addition, a significant part of cadmium can enter the soil with phosphorus fertilizers and lime materials. Furthermore, for many decades, the Aral Sea has been a sink for the particulate and dissolved matter transported by the Amudarya and Syrdarya rivers. This matter included geogenic mineral

Table 4.10 Mobile and gross forms of heavy metals in the meadow and floodplain-meadow soils (mg/kg)

	Mobile			Gross		
Soil profile depth (cm)	Zn	Cu	Cd	Zn	Cu	Cd
Meadow soils						
0–10	6.55	3.35	0	27.8	8	12.2
10–20	3.5	2.45	0	27.8	9.2	14.8
20–50	6.55	3.5	**0**	67.4	15.2	8.4
Floodplain-meadow soils						
0–10	5.07	2.11	1.02	48.15	19.7	11.2
10–20	4.12	3.1	0.75	36.7	16	12.3
20–50	10.47	1.95	1.5	62.7	14.5	14.1

loads as well as unprocessed wastewater with high concentrations of heavy metals, especially lead and cadmium (Wiggs et al. 2003), radioactive substances (uranium and radon) from the ore processing industry in the upstream countries of Tajikistan and Kyrgyzstan (Kulmatov and Hojamberdiev 2010), urban and industrial wastewater discharges from an increase in inhabitants (Dukhovny and de Schutter 2011), and highly concentrated, leached agrochemicals, pesticides, and salts used to combat widespread salinization (Micklin 2007; O'Hara et al. 2000; Whish-Wilson 2002).

4.3 Soil Texture

The soil texture has a great influence on soil formation and agricultural properties of soils. The processes of soil movement; transformation and accumulation of substances; and physical, physical-mechanical, and water properties of the soil, such as porosity, moisture capacity, water permeability, water-carrying capacity, structure, air, and thermal conditions depend on the soil texture.

The soil types encountered in the dried Aral seabed differed in soil texture (Table 4.11). The soils of the dried seabed are mainly slightly loamy (29.4%), sandy loam (26.8%), medium loam (19.2%), sands (14.9%), and clay (4.8%) (Fig. 4.22).

The soils are sandy loam in the *desert sands*; sand accounted for 83.2%, dust 13.1%, and clay 3.3.

In *desert sands with brown soil formation*, fractions of sand (69.2%), dust (20.2%), and clay amounted to 10.7%.

The soil texture of *solonchaks (typical, coastal, takyr-like, and solonchaks with saline marine sediments)* ranged from 38.3% to 49.8% as sand, dust (40.3–50.4%), and clay (7.9–13.6%).

The soil texture is loamy in the *meadow and floodplain-meadow soils*; sand fractions were from 30.4% to 51.4%, dust was 53.5% and 41.9%, and clay was 16% and 6.7%.

Table 4.11 Soil texture of the different soil types in the dried Aral seabed in Kazakhstan part

Soil types	Sand	Dust	Clay	Soil texture
Desert sands	83.21	13.13	3.31	Sandy loam
Sands with brown types of soil formation	69.2	20.17	10.68	Loamy
Typical solonchaks	41.94	44.43	13.58	Loamy
Takyr-like solonchaks	49.26	40.33	10.34	Loamy
Coastal solonchaks	49.82	42.26	7.86	Loamy
Solonchaks with saline marine sediments	38.3	50.39	11.25	Loamy
Brown soils	82.13	12.53	5.31	Sandy loam
Meadow soils	30.44	53.53	15.98	Loamy
Floodplain-meadow soils	51.39	41.9	6.65	Loamy
Gray-brown soils	41.74	49.16	9.07	Loamy

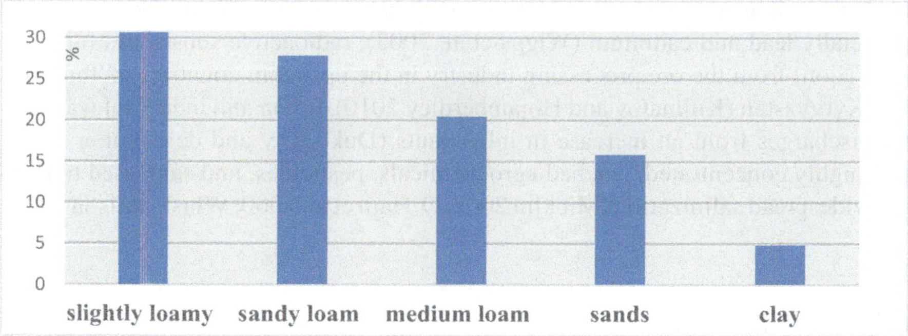

Fig. 4.22 Soil texture of the dried Aral seabed

The *brown soils* are sandy loam and sand fractions accounted for 82.1%; however, sands were 41.7% in *gray-brown soils*, dust amounting to 12.5% to 49.2%, and clay from 5.3% to 9.0%, respectively.

References

Dukhovny VA, de Schutter JLG (eds) (2011) Water in Central Asia-past present future. Balkema Press, Leiden, p 408

Faizov KSh, Talapova AS (2003) Ecology of the Crisis Territory of the Aral Sea Region and Problems of Their Solution. Almaty, 110 p (in Russian)

Kulmatov R, Hojamberdiev M (2010) Distribution of heavy metals in atmospheric air of the arid zones in Central Asia. Air Qual Atmos Health 3:183–194. https://doi.org/10.1007/s11869-010-0067-6

Micklin PP (2007) The Aral Sea disaster. Ann Rev Earth Planet Sci 35:47–72. https://doi.org/10.1146/annurev.earth.35.031306.140120

O'Hara SL, Wiggs GF, Mamedov B, Davidson G, Hubbard RB (2000) Exposure to airborne dust contaminated with pesticide in the Aral Sea region. Lancet 355(9204):627–628. https://doi.org/10.1016/S0140-6736(99)04753-4

Whish-Wilson P (2002) The Aral Sea environmental health crisis. J Rural Remote Environ Health 1(2):29–34. http://citeseerx.ist.psu.edu/viewdoc/download?doi=10.1.1.581.3798&rep=rep1&type=pdf

Wiggs GFS, O'Hara S, Wegerdt J, Van Der Meer J, Small I, Hubbard R (2003) The dynamics and characteristics of aeolian dust in dryland Central Asia: possible impacts on human exposure and respiratory health in the Aral Sea basin. Geogr J 169:142–157. https://doi.org/10.1111/1475-4959.04976

Chapter 5
Formation, Degradation, and Mapping

Abstract The new soil cover was formed due to long-term drying process and shrinkage of the Aral Sea. Saline soils in the seabed as solonchaks with several developmental stages evolutionarily transformed to sands.

Solonchaks cover 56.1%, and sands 24.39% and 19.51% belong to brown soils from the studied soils. Soil salinization is one of the main types of degradation that occurred in the dried seabed and part of the overall desertification. According to our compiled NDSI map, 26.1% of the dried seabed in Kazakhstan part was covered by extremely saline and 13.5% was strongly saline soils. Furthermore, our study and mapping results and comprehensive analysis show that the dried seabed in Kazakhstan part was degradated by 87.5% in significant degree and temperate degradation was 12%.

Keywords Aridization · Degradation · Desert · Mapping

5.1 Soil-Forming Processes

The drying processes of the Aral Sea led to the formation of a new soil cover, which is formed on the basis of exposed continental soils. It should be considered as soils that differ from zonal ones in their specific features, which consist in the dynamics of the development of the soil-forming process in space and time. This allows the newly formed soil cover of the dried seabed to go through a secular cycle of development in a short period of time. The soil-forming process begins from the moment the seabed is exposed.

The evolution of solonchaks goes through several stages: excessively hydromorphic solonchaks → moderately hydromorphic solonchaks → semihydromorphic solonchaks → semiautomorphic solonchaks → automorphic solonchaks → desert type soils (sands). At the last stages of soil development, solonchak processes caused by hydromorphic conditions die out, and the role of the arid-zonal factor increases many times over, under the influence of which further development of soils proceeds typically according to the desert type.

G. Issanova et al., *Soil Cover of the Dried Aral Seabed in Kazakhstan*,
https://doi.org/10.1007/978-3-031-29867-7_5

The hydromorphic solonchaks become dangerous when they dry out and the leaching regime of solonchaks changes to an effusion water-salt regime, and salt accumulates in the surface horizons and a salt crust forms, and under the action of the wind, the salt is carried over long distances. The most susceptible to erosion among semihydromorphic and semiautomorphic solonchaks are puffy, crusty-puffy solonchaks, the surface of which is represented by a powdery earthy-salt layer, and when the crust is destroyed, they become the focus of salt-dust transfer of the powdery puffy subcrustal layer. Such soils are a source of dust and salt carried by the wind outside the Aral Sea basin. The most dangerous are loose sands. The overgrowth of solonchaks by artificial plantings or self-overgrowth reduces the environmental hazard and significantly reduces it when solonchaks pass into desert-sandy soils.

The study of the dried seabed soils showed that the chain of solonchak transformation ends with the formation of desert-sandy soils. The parent rock of modern soil formation on the dried Aral seabed has marine, lacustrine, alluvial, and aeolian genesis.

For the period 1990–2020, there was a significant decrease in the hydromorphic solonchaks by 15.1% due to the development of the process of aridization and, accordingly, an increase in the automorphism of solonchaks, a decrease in the level of groundwater and the transition of hydromorphic soils to their automorphic counterparts, sanding, and the formation of zonal soil. There is a formation of desert-sandy soils, an increase in their area by 5.7%, which is a positive sign. The increase in the sand cover indicates the intensification of erosion processes on the dried seabed.

The soils of the Aralkum desert are saline (5–20 kg/m^3) and weakly fixed by vegetation, which are subjected to intense soil deflation. The salts contained in the soils are carried out to the surrounding areas.

5.2 Soil Types and Soil Transformation in Soil Properties and Types

The following soil types were determined: 56.1% to solonchaks (continental, coastal, takyr-like, solonchaks in places with dried marine saline sediments), 24.39% of the studied soil profiles belong to sands (desert, desert with brown-type soil formation), and 19.51% to brown soils (meadow, named-meadow, gray-brown desert, in places with a light brown).

The soil cover of the dried seabed is an indicator of ecological stability. From the point of view of soil cover, the group of ecological risk primarily includes sandy massifs: dunes, sandy hills, if they are loose or weakly fixed, and soils covered with a sand plume and saline soils (solonchaks).

Changes in Soil Properties Since 1977, the soil in the Kaskakulan site has transformed from coastal solonchaks to ordinary solonchaks. In 1977, the soils were belonged to coastal solonchaks. There were more clay fractions at the bottom of the soil profile and up to 50 cm medium saline; further to a depth of 100 cm, it was highly saline (0.35–2.96%). Humus was in very small quantities (0.03–0.44%), calcium and magnesium cations predominated in the composition of the absorbed bases, the reaction (pH) of the soil was highly alkaline, and the CO_2 content was high (Table 5.1). However, in 2020, the soil belonged to ordinary solonchaks and noticed some changes in soil properties. Silty fractions were higher on the soil surface, and the soil horizon was nonsaline (0.05–0.08%). Humus was in insignificant amounts (0.45%), calcium and magnesium cations predominated in the composition of the absorbed bases, the reaction was alkaline, and the CO_2 content was low (Table 5.1).

The soil transformation took place in the Karateren site: coastal solonchaks to solonchaks with saline sediments. The nature of the distribution of clay fractions in the soils was greater in 2020 compared to 1977. The humus content did not differ (very low). Soil salinity was more in 1977 (2.31%) over the whole soil horizon, and in the upper horizon, there was a very high distribution of salts (9.09%). In 2020, the salinity was 1.01% (whole horizon) and in the upper horizons (0–10 cm and from 30 to 100 cm) were highly saline. The amount of absorbed bases was greater in 2020; if in 1977 the amount of absorbed bases were dominated by calcium and magnesium cations, then in 2020 sodium predominates. The soil reaction was strongly alkaline in 1977 and decreased to alkaline in 2020, and CO_2 content increased by 2020 (Table 5.1).

The transformation of sands to sands with brown types of soil formation occurred in the Basay site.

In 1977, there were no silt fractions in the studied sands and the soil horizon is nonsaline; however, in 2020 silt fractions were 10.25%. The humus content was insignificant: 0.62% (1977) and 1.09% (2021). There were no absorbed bases, the reaction of the soil environment was strongly alkaline (8.5–8.8), and the average content of carbonates was low (1.83%). However, in 2021, the sums of absorbed bases were high at 24.89 mg/eq, magnesium and sodium predominated in the composition, and the reaction was weakly alkaline (7.5).

Salt Composition Changes The composition of the analyzed water extract from the *Kaskakulan site* shows the salt composition of the soil. From water extracts can be seen the nature of salt accumulation and the direction of the salinization process. According to the composition of the water extract, the following differences were obtained:

In 1977: $SO_4^{2-} > Na^{+}+K^{+} > Mg^{2+} > Cl^{-} > Ca^{2+} > HCO_3$.
In 2020: $HCO_3 > Na^{+}+K^{+} > SO_4^{2-} > Ca^{2+} > Mg^{2+} > Cl^{-} > CO_3$.

Changes in the composition of water extracts in 2020 compared to 1977 were observed. If in 1977 sulfates and sodium predominated in the composition of the water extract and the type of chemistry salinity was soda-sulfate, then in 2020 the predominant ions were bicarbonates and sodium. In 1977, the soil was highly saline, the soil surface was solonchak, and in 2020, it was nonsaline (Table 5.2). There was a decrease in soil salinity in 2020 to a nonsaline degree.

Table 5.1 Comparison of chemical parameters of the soils in the dried Aral seabed (1977 and 2020–2021)

Sites	Soil sampling depth (cm)	Clay fractions	Salinity	Humus (%)	Absorbed bases mg/eq				pH	CO_2
					Total	Ca	Mg	Na + K		
	1977 (soil profile 2)									
Kaskakulan	0–10	7.9	2.96	0.44	–	7.5	–	0.4	8.65	5.7
	10–20	3.4	0.72	0.08	7.35	3	3.5	0.85	8.21	6.4
	20–30	2.9	0.35	0.08	6.84	3	3	0.84	8.25	6.4
	30–50	4	0.4	0.05	6.81	2	4	0.81	8.27	6.4
	50–100	24.8	1.01	0.03	15.39	4.66	8.33	2.39	8.48	8.03
	2020 (soil profile 1)									
	0–10	7.212	0.068	0.45	1.56	0.96	2.45	0.28	7.68	1.49
	10–20	3.603	0.051	–	2.72	3.43	4.41	0.33	7.79	0.62
	20–30	3.603	0.051	–	2.72	3.43	4.41	0.33	7.79	0.62
	30–50	5.604	0.059	–	1.55	1.96	2.45	0.26	7.81	0.58
	50–100	4.6	0.08						7.98	0.69
Karateren	*1977 (soil profile 4)*									
	0–10	2.3	9.09	–	9.03	8.5	Нет	0.53	9	5.3
	10–20	19.5	1.23	1.02	17.24	7	9.5	0.74	8.6	9.3
	20–30	14.8	0.399	0.95	18.73	6.5	10	2.23	86	8.8
	30–50	1.6	0.568	0.32	2.58	1.5	0.5	0.58	9.2	5.6
	50–100	11.6	0.3	–	13.62	6	6.5	1.12	8.7	8.2
	2020 (soil profiles 20, 21)									
	0–10	13.17	1.26	1.42	38.67	12	12	14.66	8.02	7.61
	10–20	11.35	081	0.6	41.78	19.6	8.08	14.1	8.25	7.6
	20–30	11.13	0.53	0.5	25.58	16.66	8.33	0.59	8.21	7.64
	30–50	13.73	1.34	0.4	9.67	4.9	4.41	0.36	8.39	7.69
	50–100	9.37	1.12						8.45	7.71
Basay	*1977 (soil profile 5)*									
	0–10	–	0.19	0.62	–	–	–	–	8.5	2.76
	10–20	–	0.6	0.25	–	–	–	–	8.6	2.27
	20–30	–	0.9	0.17	–	–	–	–	8.8	2.27
	30–50	–	0.3	0.25	–	–	–	–	8.6	1.46
	50–100	–	0.52	–	–	–	–	–	8.6	1.2
	2021 (soil profile 17)									
	0–10	12.867	3.58	1.09	31.76	8.42	19.31	12.45	7.348	1.64
	10–20	7.235	1.85	0.93	35.24	11.88	5.94	17.42	7.141	2.97
	20–30	9.65	1.58	0.65	22.92	1.98	6.93	14.01	7.614	4.1
	30–50	12.475	1.61	–	25.13	3.47	7.43	14.23	7.954	4.07
	50–100	9.053	0.83	–	9.43	0.99	4.95	3.49	7.337	0.55

Table 5.2 Comparison of water extract results from the soils of the dried Aral seabed (1977 and 2020–2021)

Site	Soil sampling depth (cm)	HCO_3^-	CO_3^-	Cl^-	SO_4^{2-}	Ca^{2+}	Mg^{2+}	Na^++K^+	Total salt (%)
Kaskakulan	*1977 (soil profile 2)*								
	0–10	0.36	–	10	35.6	11.1	27.9	6.9	2.955
	10–20	0.14	–	6.6	4.9	0.7	2.5	8.4	0.719
	20–30	0.16	–	3.4	2	0.5	0.9	4.1	0.346
	30–50	0.16	–	3.6	2.6	0.6	1.1	4.6	0.4
	50–100	0.28	–	11.82	6.42	1.47	3.75	13.25	1.12
	2020 (soil profile 1)								
	0–10	0.48		0.04	0.37	0.3	0.2	0.39	0.068
	10–20	0.44	0.04	0.08	0.14	0.2	0.1	0.35	0.051
	20–30	0.44	0.04	0.08	0.14	0.2	0.1	0.35	0.051
	30–50	0.48	0.04	0.08	0.21	0.3	0.1	0.37	0.059
	50–100	0.52	0.04	0.13	0.46	0.25	0.15	0.71	0.08
Karateren	*1977 (soil profile 4)*								
	0–10	0.54	0.03	97.8	52.1	12.55	33.1	104.8	9.09
	10–20	0.36	–	9.6	9.5	5	2.5	11.9	1.23
	20–30	0.48	–	4.9	3.4	1.4	1.2	6.2	0.399
	30–50	0.61	–	5.2	3.1	1.1	1.2	6.6	0.568
	50–100	0.62	–	2.2	1.85	0.9	0.7	3.05	0.3
	2020 (soil profile 20, 21)								
	0–10	0.26	0	7.43	12.25	3.46	5.45	11.04	1.26
	10–20	0.34	0.04	5.23	7.4	1.73	3.71	7.52	0.81
	20–30	0.48	0.04	1.21	6.31	5.45	3.47	1.57	0.53
	30–50	0.48	0	3.69	16.35	9.16	4.7	6.41	1.34
	50–100	0.38	0	3.96	12.91	4.7	5.57	7.72	1.12
Basay	*1977 (soil profile 5)*								
	0–10	0.33	–	0.14	0.1	0.4	0.08	0.09	0.19
	10–20	0.28	–	0.11	0.13	0.35	0.98	0.09	0.36
	20–30	0.28	–	0.11	0.13	0.35	0.98	0.09	0.09
	30–50	0.33	–	0.11	0.13	0.4	0.08	0.09	0.1
	50–100	0.25	–	0.23	2.38	2.35	0.25	0.26	0.52
	2021 (soil profile 17)								
	0–10	0.4	0.2	41.6	22.8	22.6	15.6	11.3	3.58
	10–20	0.35	–	13.2	15.1	6.6	2	20.39	1.85
	20–30	0.4	–	16.4	8.5	1.8	2.8	20.96	1.58
	30–50	0.5	–	14.4	10.7	2.6	2.8	20.00	1.61
	50–100	0.45	–	7.2	5.4	1.2	1.2	10.93	0.83

The composition of the water extract from the *Karateren* is arranged in the following descending rows:

In 1977: Na^{+}+K^{+} > Cl^{-} > SO_4^{2-} > Mg^{2+} > Ca^{2+} > HCO_3 > CO_3.
In 2020: SO_4^{2-} > Na^{+}+K^{+} > Mg^{2+} > Ca^{2+} > Cl^{-} > HCO_3 > CO_3.

The following changes in the composition of water extracts were revealed: in 1977, sodium and chlorides dominated in the composition; the sodium content was very high (104.8 ml/eq), which increased the degree of salinity of this horizon to 9.09%. In addition, there were high levels of chlorides (97.8%) and sulfates (52.1%). The type of soil salinization chemistry was chloride-soda. The 10–20 cm horizon was highly saline and the remaining horizons were moderately saline. Intensive salinization of the soil surface was observed, that is, initial soil salinity.

In 2020, in this area, the composition of the water extract is slightly different; however, the salinity of the soil is not strong. Sulfate and sodium ions predominated in the aqueous extract. The sodium values were not as high as in 1977. In general, the soil profile was highly and moderately saline.

According to the water extract of *Basay site*, the following differences were obtained:

In 1977: Ca^{2+} > SO_4^{2-} > Mg^{2+} > HCO_3 > Cl^{-} > Na^{+}+K^{+}.
In 2021: Cl^{-} > Na^{+}+K^{+} > SO_4^{2-} > Ca^{2+} > Mg^{2+} > HCO_3 > CO_3.

In 1977, the predominant ions in the water extract were calcium and sulfates. In general, nonsaline sands are up to half a meter, and slightly saline by 10–20 cm, moderately saline from 50 to 100 cm. In 2021, chloride and sodium were the predominant ions. In terms of salinity, the upper 0–10 cm horizon is solonchak (3.58%), and the remaining soil horizons are highly saline (>1%). The type of soil chemistry is soda-chloride.

We can say that there is an intensive soil salinization of all horizons and the distribution of salts is uniform along the soil profile.

Transformation in Soil Types Since the 1960s, the retreat of the Aral Sea began. According to the results on soil researches conducted in the northeastern part of the dried seabed in 2020–2022, during 1960s–1980s, solonchaks turned into solonetzs soils. In the eastern part toward the south closer to the coastlines, takyr-like soils are common and further toward the sea, continental solonchaks are developed and sands in small quantities to the south (Fig. 5.1). Desert sands, continental solonchaks, and solonchaks with saline sediments are widespread distributed along the coastlines in the southeastern part.

During 1990–2000, there was a deep drying of the Aral Sea. In the 1990s, in the eastern part of the dried seabed, solonchaks with saline sediments were distributed and in the southeastern part, solonchaks with saline sediments and sands. In 2000s, from the east to the southeast, solonchaks with saline sediments are distributed, and closer to the sea, strongly degraded coastal solonchaks are widespread (Fig. 5.1). In the central part of the dried seabed, gray-brown desert soils and light brown soils are common.

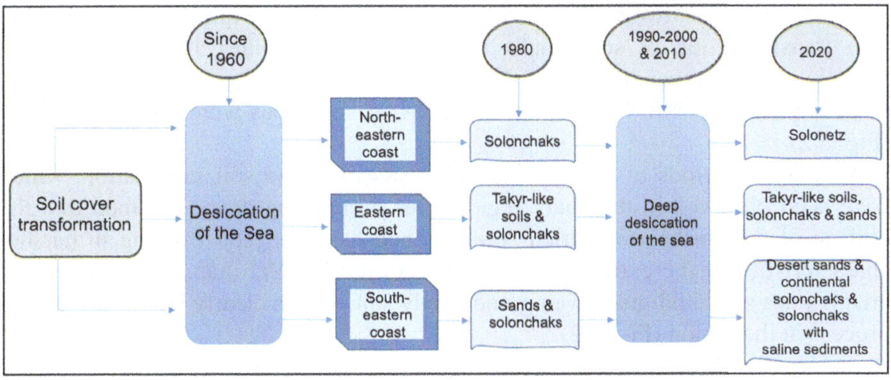

Fig. 5.1 Long-term soil transformation process in the dried Aral seabed

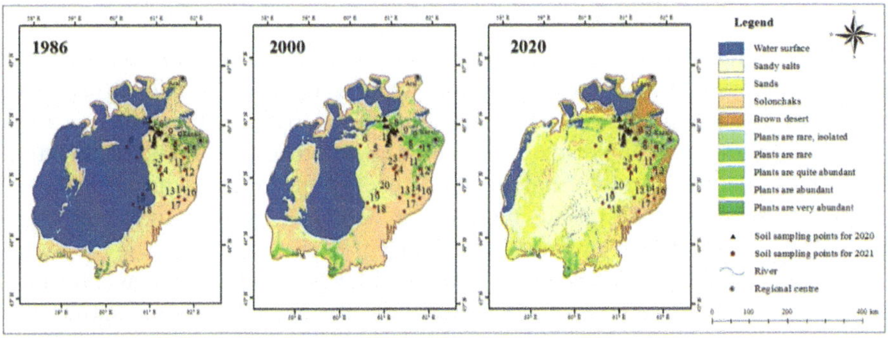

Fig. 5.2 MSAVI map of the dried Aral seabed for 1986, 2000, and 2020

In the 2010s, there was an increase in the dried territory of the Aral Sea. Strongly degraded coastal solonchaks and further solonchaks with saline sediments and sands are common in exposed seabed. To the east, there is a small island with gray-brown desert soils and with light brown soils.

Furthermore, the Modified Soil-Adjusted Vegetation Index (MSAVI) was calculated to determine the soil characteristics and update the soil map (Fig. 5.2). Data obtained as a result of fieldwork and satellite images such as Landsat 8 in 2020 were used to analyze changes in land use/vegetation.

The results of the updating of the soil map using data from fieldworks in 2020 and MSAVI showed that the floodplain-meadow soils along the Syrdarya river extended to the Aral Sea (17 and 18 soil profiles). Sands (soil profile 6), meadow soils (soil profile 16), meadow swamps (soil profile 19), and takyrs (soil profiles 2, 4, 5, 14) were formed on the territory of solonchaks. Brown desert soils were formed on the territory of floodplain-meadow soils (soil profile 9).

According to fieldwork data in 2021, in the central part of the eastern coast, there is a transformation of solonchaks into sands (soil profiles 1–3). The area of Kyzylkum sands expanded in a northwesterly direction (soil profiles 11–13, 16, 17). In general, the spatial distribution and areas of soil types were changed slightly (Fig. 5.2).

Prolonged periods of drought lead to desiccation and soil salinization, which have been observed in the study region. Our monitoring data confirmed that the climate of the northeastern dried Aral Sea bed has changed, resulting in harsher characteristics of the extremely arid zone. On the MSAVI map, the predominant area is pure sand and mud/swamp and sandy salt, which clearly shows the drying process in the region (Fig. 5.2).

5.3 Soil Degradation Processes

Soil degradation as a salinization have occurred in arid and semiarid areas, including Aral Sea region as a result of various factors, including irrational use water resources and human activities. They are part of the overall desertification processes.

Soil salinization is one of the most well-known processes of land degradation. Anthropogenic impact on the soil cover and ecosystems of Kazakhstan in all geographical areas, mostly Aral Sea region, increases year by year. Irrational use of natural resources leads to environmental damage.

In Central Asia, the degradation of irrigated lands mainly occurs as a result of secondary soil salinization, irrigation, and wind erosion, a decrease in humus content and organic matter, soil pollution with agrochemicals, waterlogging of lands, and the development of agricultural pests. Almost 60% of the population of Central Asia depends on agriculture as a source of food and income (Qadir et al. 2009). At the same time, from 40% to 60% of irrigated lands in the region are saline or waterlogged. Soils are most affected in Turkmenistan (68% of the total area), Uzbekistan (52%), and Kazakhstan (41%) (FAO 2020).

According to rough estimates, the area of saline soils in Kazakhstan is more than 110 million ha (41% of the entire territory of the country). They are mainly widespread in the southern and central regions of Kazakhstan. In these dry areas, annual precipitation is 100–150 mm and evaporation exceed (Issanova et al. 2017). Only in Kyzylorda region, more than 73,000 ha of irrigated land are subject to severe soil salinization. The soils of the region are moderate and highly saline. The region is suffering most from the effects of the dried Aral Sea (FAO 2020).

The Aral Sea is an ecological catastrophic crisis, which occurred since the 1960s and is expressed in the most acute social and environmental problems, including soil salinization and desertification, degradation of vegetation (Issanova et al. 2022), climate change (Lioubimtseva 2014), progression of various diseases among population of the region (Anchita et al. 2021), and other negative processes. As a result of sea drying, the Aralkum sandy-salt desert formed with almost 4 million ha territory in the former seabed and only halophytes can be grown in this area (Matzhanova

2007). The resulting desert intensifies the processes of desertification of landscapes around the former sea because the rapid drying of the Aral Sea has led to global changes in all natural conditions, including changes in soil cover (Faizov and Tapalova, 2003; Stulina and Sektimenko 2004; Issanova et al. 2022) and climate (Lioubimtseva 2014). Since soil is an indicator of natural processes, and its condition is the result of long-term exposure to various sources of pollution and deterioration, the process of primitive soil formation in the seabed began on various sandy deposits, sandy loamy rocks exposed on a dried bottom (Orazbekova and Budnikova 2014). These deposits with a light-textured composition led to an increase in deflationary processes in the region. As a result, saline dust was introduced into the Aral Sea region, including irrigated lands. Soil deflation process as dust and salt storms was noted with periodic removal of sandy and salt materials from all types of solonchaks in the dried seabed (Issanova et al. 2015; Zucca et al. 2021). Dust storms, droughts, and dry and strong winds are the most significant climatic factors of desertification (Issanova et al. 2014; Semenov 2011). According to the national report, 75 million tons of sand, as well as 65 million tons fine dust and salt, rise from the dried bottom of the Aral Sea, where winds and cyclones constantly blow. This leads to a decrease in pasture productivity and crop yields. The irrigated lands of the lower reaches of the Amudarya and Syrdarya annually receive 100 million tons of salt and sand (Nauryzbayeva 2019; Aiimbetova 2020; Ramazonov et al. 2020).

Soil salinity is one of the main characteristics of soils in arid regions, limiting soil fertility and determining their genetic and meliorative properties. It quickly responds to changes in natural conditions and anthropogenic impacts and is an informative property of the state of soils. Therefore, in arid regions, it is monitoring of changes in soil salinity that makes it possible to obtain the necessary information about the processes taking place in soils.

According to the compiled and updated soil map and its degradation, the dried seabed is covered mainly by solonchaks (continental and seaside) that undergo anthropogenic salinization: desert sands, gray-brown desert and brown desert solonetzic soils, takyrs and takyr-like soils, and meadow boggy. Since the 1960s, the Aral Sea has been dried up, and only 10% of the seawater was left (Wu et al. 2021) and the newly formed saline surface as a desert has been found to be an active and potential source of aeolian sand, salt, and dust (Issanova et al. 2015; Shen et al. 2016). The former seabed is now a saline and sandy anthropogenic desert called Aralkum because the inflow water from the Syrdarya and Amudarya rivers to the sea was used constantly and irrationally mainly for irrigation.

According to our compiled NDSI map, 26.1% of the dried seabed in Kazakhstan part was covered by extremely saline and 13.5% was strongly saline soils (Fig. 5.3). The highest salinity values were found on the dried bed of the Aral Sea toward the central part of the bed (Fig. 5.3). Results of lab analysis and NDSI mapping show strong visual correlation. Salt content of the studied soils matches with mapping results.

In addition, our study and mapping results and comprehensive analysis show that the dried seabed in Kazakhstan part was degradated by 87.5% in significant degree and temperate degradation was 12% (Issanova et al. 2022). Degradation types as an

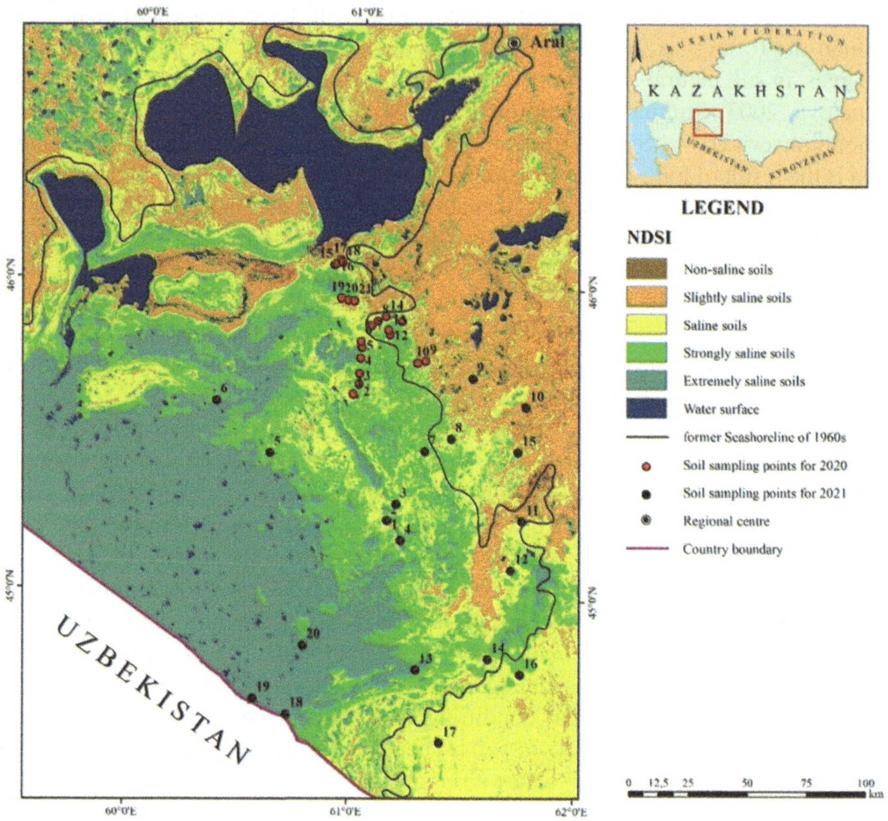

Fig. 5.3 NDSI map of the dried Aral seabed

anthropogenic soil salinization, soil alkalinization, and wind erosion prevail in the dried seabed. In addition, degradation has environmental impacts on vegetation cover and water quality in the forms of pollution and deterioration.

5.4 Soil Maps of the Seabed (1980s and 2020s)

The soil map of the eastern dried Aral seabed for the 80s was compiled on the basis of the soil map of Kazakhstan in 1976 at a scale of 1:2,500,000. To compare the reliability of GIS methods for our further research, satellite imagery data were added to the existing retrospective data, since satellite images for the 80s do not provide a complete picture of the soil condition.

As a result of comparing the soil map for 1975 and satellite images (Landsat 8) for 1986, we notice that the eastern coastline has receded 20–60 km to the west. According to satellite images, peculiar processes of soil formation took place on the

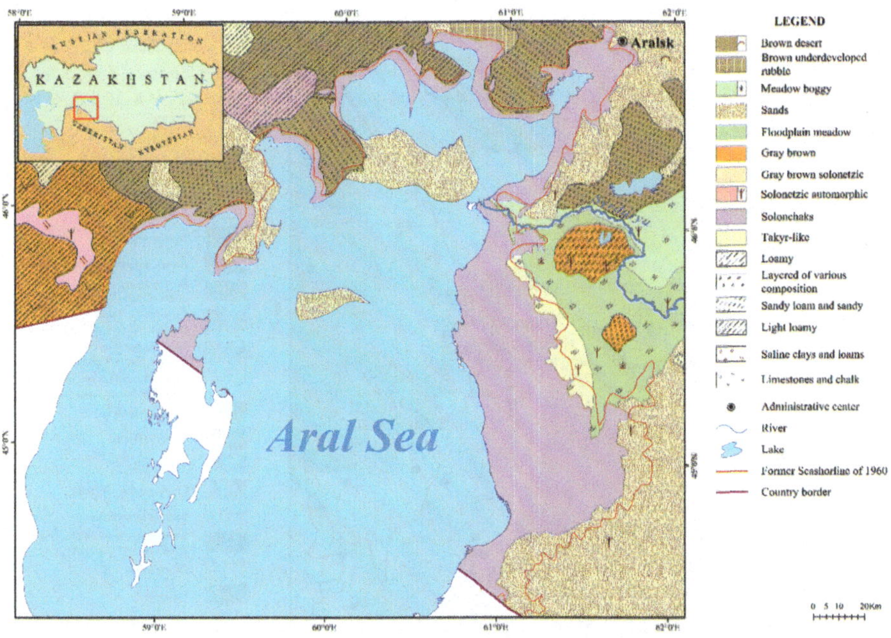

Fig. 5.4 Soil map of the dried Aral seabed for the 1980s

dried part (Fig. 5.4). Exposed seabed sediments are deposits of marine, river, and mixed genesis. The soil cover of the dried Aral seabed goes through a secular cycle of development in a short period of time. The initial stage of the formation of the soil cover is associated with the intensive development of salt accumulation on the soils released from under the water, and this is confirmed by the Landsat 4–5 satellite images for 1986. As a result of deciphering old satellite images and studying the soil of the 1980s, a soil map of the southeastern coast of the dried bed of the Aral Sea was compiled on a scale of 1:200,000 (Fig. 5.4).

The soil map of the dried bed of the Aral Sea for 2020s was compiled on the basis of existing archive cartographic and literary materials from previous years of research. The existing cartographic materials, fieldwork data, and satellite images were systematized and summarized in order to make changes on it and update the soil map (Fig. 5.5). Cartographic materials were processed using ArcGIS® software.

The existing soil map of Kazakhstan at a scale of 1:2,500,000 taken from National Atlas of the Republic of Kazakhstan 2010 was used as the basis for the soil map. Geographically referenced points of soil profiles were superimposed in a layer on the digitized soil map.

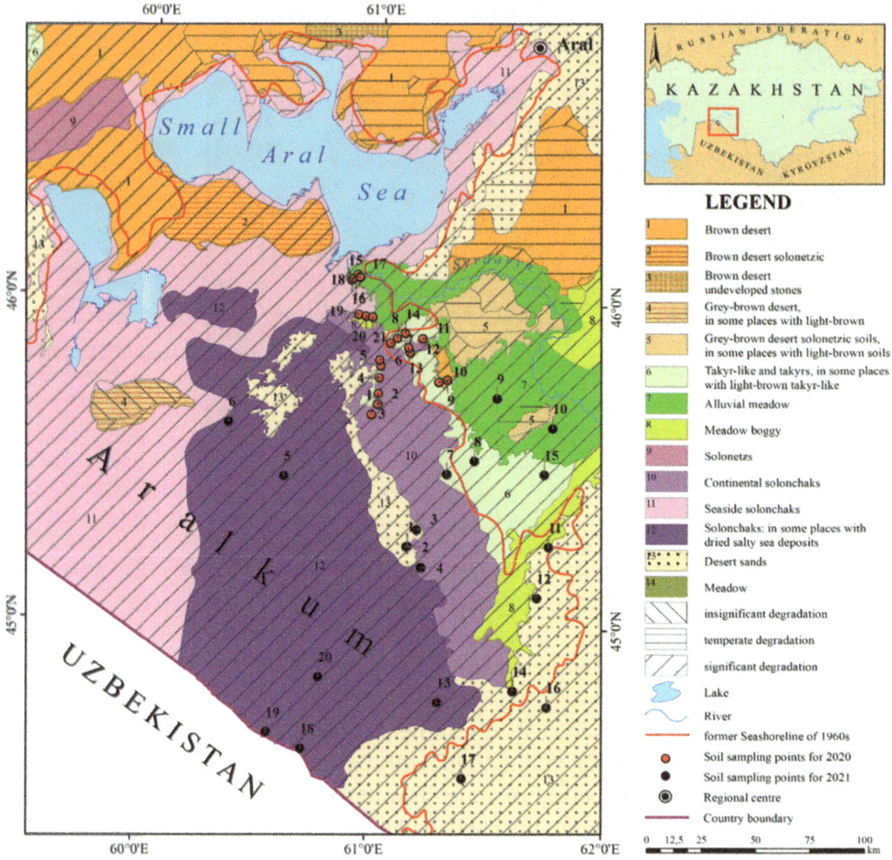

Fig. 5.5 Soil map of the dried Aral seabed for 2020s

References

Aiimbetova S (2020) Environmental monitoring of soil salinization in the southern Aral Sea region. Eurasian Union Sci 9–78:10–12. (in Russian)

Anchita ZA, Khaibullina Z, Kabiyev Y, Persson K, Tussupova K (2021) Health impact of drying Aral Sea: one health and socio-economical approach. Water 13:3196. https://doi.org/10.3390/w13223196

Faizov KSh, Talapova AS (2003) Ecology of the crisis territory of the Aral Sea region and problems of their solution. Almaty, p 110. (in Russian)

FAO (2020) Capacity development program in Central Asia on soil salinity mapping and soil fertility restoration. Global Soil Partnership. https://www.fao.org/global-soil-partnership/resources/highlights/detail/en/c/1306719/

Issanova G, Abuduwaili J, Semenov O (2014) Deflation processes and their role in desertification of the southern pre-Balkhash deserts. Arab J Geosci 7:4513–4521

Issanova G, Abuduwaili J, Galayeva O, Semenov O, Bazarbayeva T (2015) Aeolian transportation of sand and dust in the Aral Sea region. Int J Sci Technol 12:3213–3224

Issanova GT, Abuduwaili J, Mamutov ZU, Kaldybaev AA, Saparov GA, Bazarbaeva TA (2017) Saline soils and identification of salt accumulation provinces in Kazakhstan. Arid Ecosyst 7(4):243–250

Issanova G, Abuduwaili J, Tynybayeva K, Kalybayeva A, Saduakhas A, Kulymbet K, Kaldybayev A, Erlan G, Tanirbergenov S (2022) Soil salinisation as a land degradation process in the dried bed of the North-Eastern Aral Sea, Kazakhstan. Arab J Geosci 15:1055. https://doi.org/10.1007/s12517-022-09627-w

Lioubimtseva E (2014) Impact of climate change on the Aral Sea and its basin. In: Micklin P et al (eds) The Aral Sea, Springer earth system sciences. Springer, Berlin/Heidelberg. https://doi.org/10.1007/978-3-642-02356-9_17

Matzhanova (2007) Haloadaptation and ecological characteristics of plants in the southern dried bottom of the Aral Sea part due to desertification and soil salinization. Arid Ecosyst 13(33–34):113–117. (in Russian)

Nauryzbayeva Z (2019) Problems of soil formation and salinization in the southern Aral Sea region. Eurasian Union Sci 9–66:4–5. (in Russian)

Orazbekova K, Budnikova T (2014) Environmental assessment of natural complexes of the dried bottom of the Aral Sea in the territory of Kazakhstan for the purpose of prospective development. Eurasian Union Sci 5:48–52. (in Russian)

Qadir M, Noble AD, Qureshi AS, Gupta RK, Yuldashev T, Karimov A (2009) Salt-induced land and water degradation in the Aral Sea basin: a challenge to sustainable agriculture in Central Asia. Nat Res Forum 33:134–149

Ramazonov B, Faiziev V, Mutalov K, Zakirov D (2020) Desertification and transformation soil cover in the Aral Sea region. Acad Res Educ Sci 1(2):213–220. (in Russian)

Semenov OE (2011) Introduction to experimental meteorology and climatology of the sand storms, Almaty, p 580. (in Russian)

Shen H, Abuduwaili J, Samat A, Ma L (2016) A review on the research of modern aeolian dust in Central Asia. Arab J Geosci 9:625

Stulina G, Sektimenko V (2004) The change in soil cover on the exposed bed of the Aral Sea. J Mar Syst 47:121–125

Wu Q, Yue H, Liu Y, Hou E (2021) Geospatial quantitative analysis of the Aral Sea shoreline changes using RS and GIS techniques. Earth Sci Inf. https://doi.org/10.1007/s12145-021-00714-2

Zucca C, Middleton N, Kang U, Liniger H (2021) Shrinking water bodies as hotspots of sand and dust storms: the role of land degradation and sustainable soil and water management. Catena 207:105669. https://doi.org/10.1016/j.catena.2021.105669

Chapter 6
Conclusions

Abstract Since the 1960s, as a result of various factors, including irrational use water resources from Syrdarya and Amudarya rivers, and human activities, soil degradation as a salinization and deflation have occurred in the exposing bed of the Aral Sea. They are part of the overall desertification processes.

Since the 1960s, as a result of various factors, including irrational use water resources from Syrdarya and Amudarya rivers, and human activities, soil degradation as a salinization and deflation have occurred in the exposing bed of the Aral Sea. They are part of the overall desertification processes.

The soil cover in the dried Aral seabed is covered by saline soils (different types of solonchaks) and sands. In addition, according to the results of research and mapping, meadow and floodplain-meadow soils, brown desert and alkaline soils, and desert sands on the northern and northeastern coasts of the dried Aral seabed in the Kazakhstan part were identified.

The soil properties of the subtypes of the main soil types do not differ much from each other. For instance, sands are characterized by low humus content, low carbonate content, and slightly alkaline reaction of the soil environment.

Solonchaks also have low humus content (below 1%) and high carbonate content (7–10% CO_2) typical for soils in desert zones. Alkalinity of the soil environment was pH 7–8 and with high absorption capacity (12–40). In terms of salt content, all subtypes of solonchaks are strongly saline and highly saline; the amount of salts ranges from 1.34% to 16%. In the composition of water extracts from all solonchaks, high contents of sodium, chlorides, and sulfates were noted. Solonchaks with saline sediments were distinguished by very high salt content.

In brown types, low humus content and high carbonate content was up to 21%, and alkalinity of the soil was 7.5–7.9. Various values of absorption capacity are also noted. The brown type turned out to be nonsaline, and the gray-brown subtypes were strongly saline, with a salt content of more than 2%. Sodium predominated in the aqueous extract.

Meadow soils were characterized by low humus content, medium carbonate content (5–6%), and various values of absorption capacity. In these soils, salinity ranged from medium salinity to strongly saline degree. Chloride and sodium were predominantly found in aqueous extracts.

G. Issanova et al., *Soil Cover of the Dried Aral Seabed in Kazakhstan*,
https://doi.org/10.1007/978-3-031-29867-7_6

The exposed seafloor is now a saline and sandy anthropogenic desert called Aralkum and is a powerful source of aerosols that can seriously pollute air/water, leading to anthropogenic soil salinization and vegetation degradation; furthermore, these processes greatly affect the climatic and environmental conditions in Central Asian countries and across the world. The soils in the dried seabed are contaminated mainly with heavy metals such as Cd, Zn, and Pb that can lead to soil and vegetation degradation.

The current state of the soil cover of the dried seabed is due to the process of aridization against the background of an acute shortage of water resources, mainly river inflows. As a result, soils and landscapes began to dry out and become deserted; soil salinization and deflation processes as salty storms intensified in the dried seabed.

These research results will make it possible to obtain a current soil cover data and information on the soil transformation and they can be used in compiling of soil contamination map of the dried seabed and its chemistry. In addition, the obtained research results can be useful for scientists, farmers, and decision-makers with this valuable latest information on the state of soil cover and the environment in the Aral Sea region.

Chapter 7
Recommendations for the Rational Use of the Land and Water Resources in the Dried Aral Seabed Region

Abstract According to the catastrophic situation in the Aral Sea region, it is recommended to carry out the following measures for the rational use of the land and water resources in the dried Aral seabed region:

1. Rational use of water resources, stabilization, and preservation of the sea level at the current level. Develop an international concept for the transition of agriculture in the region to the use of modern resource-saving technologies (drip irrigation, drainage systems, selection of crops that are less demanding on moisture, a fertilizer system with elements of melioration and restoration of soil fertility, laying forest plantations).
2. Phased implementation of complex land reclamation measures according to the degree of soil salinity. On saline areas of the dried seabed, carry out chemical reclamation according to the type of salinity (chloride-sulfate, sulfate-chloride, chloride-soda, sulfate-soda, soda-sulfate, and soda-chloride) followed by phytomelioration (selection of vegetation; use the type of trees in as biological pumps to lower elevated groundwater levels).
3. Preserve and increase the areas and types of existing natural vegetation cover on various soils.
4. Carrying out annual monitoring of the state of soil and vegetation cover, microclimate (installation of modern weather stations), and assessment of the impact of ongoing activities (in case of inefficiency, change the approach).

Promotion of rational use of water and land resources; involvement of the local population, farmers, public organizations, local executive bodies, international organizations, and scientific communities to preserve the Aral Sea and solve their problems.

G. Issanova et al., *Soil Cover of the Dried Aral Seabed in Kazakhstan*,
https://doi.org/10.1007/978-3-031-29867-7_7

According to the catastrophic situation in the Aral Sea region, it is recommended to carry out the following measures for the rational use of the land and water resources in the dried Aral seabed region:

5. Rational use of water resources, stabilization, and preservation of the sea level at the current level. Develop an international concept for the transition of agriculture in the region to the use of modern resource-saving technologies (drip irrigation, drainage systems, selection of crops that are less demanding on moisture, a fertilizer system with elements of melioration and restoration of soil fertility, laying forest plantations).
6. Phased implementation of complex land reclamation measures according to the degree of soil salinity. On saline areas of the dried seabed, carry out chemical reclamation according to the type of salinity (chloride-sulfate, sulfate-chloride, chloride-soda, sulfate-soda, soda-sulfate, and soda-chloride) followed by phytomelioration (selection of vegetation; use the type of trees in as biological pumps to lower elevated groundwater levels).
7. Preserve and increase the areas and types of existing natural vegetation cover on various soils.
8. Carrying out annual monitoring of the state of soil and vegetation cover, microclimate (installation of modern weather stations), and assessment of the impact of ongoing activities (in case of inefficiency, change the approach).
9. Promotion of rational use of water and land resources; involvement of the local population, farmers, public organizations, local executive bodies, international organizations, and scientific communities to preserve the Aral Sea and solve their problems.

Index

G. Issanova et al., *Soil Cover of the Dried Aral Seabed in Kazakhstan*,
https://doi.org/10.1007/978-3-031-29867-7

The manufacturer's authorised representative in the EU is Springer
Nature Customer Service Centre GmbH, Europaplatz 3, 69115 Heidelberg,
Germany. If you have any concerns regarding our products, please
contact ProductSafety@springernature.com

Printed and bound by CPI Group (UK) Ltd, Croydon, CR0 4YY
01/10/2026
02225875-0001